Genomic Imprinting

ADVANCES IN EXPERIMENTAL MEDICINE AND BIOLOGY

Editorial Board:
NATHAN BACK, *State University of New York at Buffalo*
IRUN R. COHEN, *The Weizmann Institute of Science*
ABEL LAJTHA, *N.S. Kline Institute for Psychiatric Research*
JOHN D. LAMBRIS, *University of Pennsylvania*
RODOLFO PAOLETTI, *University of Milan*

Recent Volumes in this Series

Volume 618
HYPOXIA AND THE CIRCULATION
Edited by Robert H. Roach, Peter Hackett, and Peter D. Wagner

Volume 619
CYANOBACTERIAL HARMFUL ALGAL BLOOMS: STATE OF THE SCIENCE
AND RESEARCH NEEDS
Edited by H. Kenneth Hudnell

Volume 620
BIO-APPLICATIONS OF NANOPARTICLES
Edited by Warren C.W. Chan

Volume 621
AXON GROWTH AND GUIDANCE
Edited by Dominique Bagnard

Volume 622
OVARIAN CANCER
Edited by George Coukos, Andrew Berchuck, and Robert Ozols

Volume 623
ALTERNATIVE SPLICING IN THE POSTGENOMIC ERA
Edited by Benjamin J. Blencowe and Brenton R. Graveley

Volume 624
SUNLIGHT, VITAMIN D AND SKIN CANCER
Edited by Jörg Reichrath

Volume 625
DRUG TARGETS IN KINETOPLASTID PARASITES
Edited by Hemanta K. Majumder

Volume 626
GENOMIC IMPRINTING
Edited by Jon F. Wilkins

A Continuation Order Plan is available for this series. A continuation order will bring delivery of each new volume immediately upon publication. Volumes are billed only upon actual shipment. For further information please contact the publisher.

Genomic Imprinting

Edited by
Jon F. Wilkins, Ph.D.
Santa Fe Institute, Santa Fe, New Mexico, U.S.A.

Springer Science+Business Media, LLC
Landes Bioscience

Springer Science+Business Media, LLC
Landes Bioscience

Copyright ©2008 Landes Bioscience and Springer Science+Business Media, LLC

All rights reserved.
No part of this book may be reproduced or transmitted in any form or by any means, electronic or mechanical, including photocopy, recording, or any information storage and retrieval system, without permission in writing from the publisher, with the exception of any material supplied specifically for the purpose of being entered and executed on a computer system; for exclusive use by the Purchaser of the work.

Printed in the U.S.A.

Springer Science+Business Media, LLC, 233 Spring Street, New York, New York 10013, U.S.A.
http://www.springer.com

Please address all inquiries to the publishers:
Landes Bioscience, 1002 West Avenue, 2nd Floor, Austin, Texas 78701, U.S.A.
Phone: 512/ 637 6050; FAX: 512/ 637 6079
http://www.landesbioscience.com

Genomic Imprinting edited by Jon F. Wilkins, Landes Bioscience / Springer Science+Business Media, LLC dual imprint / Springer series: Advances in Experimental Medicine and Biology

ISBN: 978-0-387-77575-3

While the authors, editors and publisher believe that drug selection and dosage and the specifications and usage of equipment and devices, as set forth in this book, are in accord with current recommendations and practice at the time of publication, they make no warranty, expressed or implied, with respect to material described in this book. In view of the ongoing research, equipment development, changes in governmental regulations and the rapid accumulation of information relating to the biomedical sciences, the reader is urged to carefully review and evaluate the information provided herein.

Library of Congress Cataloging-in-Publication Data

Genomic imprinting / edited by Jon F. Wilkins.
    p. ; cm. -- (Advances in experimental medicine and biology ; v. 626)
Includes bibliographical references and index.
ISBN 978-0-387-77575-3
1. Genomic imprinting. I. Wilkins, Jon F. II. Series.
[DNLM: 1. Genomic Imprinting. QU 475 G3345 2008]
QH450.G4662 2008
572.8'65--dc22
                                2007050095

DEDICATION

For my wife, Lizzie, and for Dash, our arena of genetic conflict

ABOUT THE EDITOR...

JON F. WILKINS is a Professor at the Santa Fe Institute in Santa Fe, New Mexico. His research interests in the area of genomic imprinting include the mechanisms underlying imprinted gene expression and the evolution of imprinted gene effects on adult behavior. His other research interests include population genetics, human demographic history, and the evolution of communication systems, including human language. Before coming to the Santa Fe Institute, Dr. Wilkins received his Ph.D. in Biophysics from Harvard University and was a Junior Fellow in the Harvard Society of Fellows. Photo used with permission from the Santa Fe Institute.

PREFACE

Genomic imprinting refers to a recently discovered phenomenon in which the expression pattern of an allele depends on whether that allele was inherited from the mother or the father. This difference in expression strategy correlates with differences in the epigenetic state of the two alleles. These epigenetic differences include DNA methylation at CpG dinucleotides, as well as modifications on the histones associated with the locus. In the simplest possible cases, the promoter region of the imprinted gene is methylated during oogenesis, but not spermatogenesis (or vice versa). This methylation (and its accompanying histone modifications) results in inactivation of the modified allele. Of course, most imprinted genes do not fall into this simplest case.

The goal of this book is neither to provide a basic introduction to imprinting, nor to provide a comprehensive survey of the current state of the field (which would necessarily span multiple books). Rather, the book covers on some of the more recent advances, with the goal of drawing attention to some of the emerging subtleties and complexities associated with imprinted genes. I hope that this will help to focus future research on these less understood aspects of the phenomenon.

The discovery of the first imprinted genes was precipitated by the discovery that the maternally and paternally derived genomes are not equivalent in mammals. Nuclear transplant experiments showed that gynogenetic embryos (in which both copies of each gene are derived from a female) do not develop into viable offspring. Androgenetic embryos (with two paternally derived genomes) are similarly inviable. In mammals, successful development requires both maternally and paternally derived alleles.

Given the existence of imprinted genes, the failure of gynogenetic and androgenetic embryos is not surprising. Approximately 100 imprinted genes have been discovered to date. A normal, biparental embryo will have a single active allele at each of these imprinted loci. A gynogenetic or androgenetic embryo will have no active expression from (approximately) half of these genes, and twice the normal expression from the rest.

The requirement for maternal and paternal genomes is just one indication of the systemic fragility that has resulted from the evolution of imprinted gene expression. At an imprinted locus, a deleterious mutation that would normally be recessive may be exposed to selection. This applies both to inherited and somatic mutations. In fact, a number of cancers have been associated either with somatic mutations affecting the active allele at an imprinted locus, or with somatic loss of imprinting, resulting in the inappropriate reactivation of the normally silenced allele.

The various clinical disorders associated with imprinting—as well as the mechanistic complexity involved in the establishment, maintenance, reprogramming, and interpretation of imprinted gene expression—mean that imprinting poses an interesting set of questions for a broad array of biologists. For some of these questions, we already have a reasonable idea of the answers; for others, we are just beginning to know how to formulate the questions. The goal of this book is to provide a rough sketch of what those answers are starting to look like, and, perhaps more importantly, to focus some attention on those questions that we will need to start pursuing in the future.

Why Do We Have Imprinted Genes?

The most prominent evolutionary explanation for the origin of genomic imprinting is the Kinship Theory of Imprinting. According to this theory, imprinted gene expression represents the outcome of an evolutionary conflict between the maternally and paternally derived alleles within an organism. More specifically, if we think of natural selection at the level of the gene, an allele's optimal strategy is actually to take on two different conditional strategies—one when the allele has been inherited from a male, and another when it has come from a female.

This selective asymmetry between maternally and paternally derived alleles has been most thoroughly understood in the context of genes that influence the distribution of maternal resources to offspring. In particular, most imprinted genes are expressed in the fetus or placenta during pregnancy and have an effect on fetal growth. For these genes, an evolutionarily stable strategy will balance a trade-off—between the benefits of acquiring more resources for the organism in which the genes are being expressed—and the indirect cost of taking resources away from the mother's other offspring, some of whom will have inherited an identical copy of the gene.

Given the possibility that a mother's offspring may have different fathers (even within litters in some species), a maternally derived allele is more likely to be found in those other offspring than is a paternally derived allele. The inclusive fitness of the paternally derived allele is therefore less affected by the indirect costs to the other offspring, with the result that the paternally derived copy will favor placing a higher resource demand on the mother than will the maternally derived copy.

The Kinship Theory has proven quite successful at explaining the growth-related effects of many imprinted genes. Whether or not the various extensions of the theory will be as successful in explaining other imprinted gene effects—some of which are covered in these chapters—remains to be seen.

The Future of Genomic Imprinting Research

The study of genomic imprinting has progressed to the point where the fact that alleles take on parental-origin-specific strategies is no longer surprising. However, new surprises have arisen as we have made progress in understanding both the causes and consequences of imprinted gene expression.

Recent studies have revealed greater and greater degrees of complexity associated with the regulation of imprinted gene expression. Many imprinted loci produce multiple overlapping transcripts. Some of these RNA transcripts may produce different splice variants of the same protein. Others may be untranslated, serving a cis-acting regulatory function, or be processed into small RNA products that serve some other (typically unknown) function.

The regulation of these various transcripts is deeply interconnected within any given cluster of imprinted genes, often resulting from a combination of maternal and paternal epigenetic modifications. Furthermore, the epigenetic modifications at imprinted loci are remodeled throughout development, leading to tissue-specific imprinting for many genes. The underlying mechanisms are just now beginning to be unraveled, and we will soon be in a position to understand the evolutionary causes and physiological consequences of this dynamic process.

Most imprinted genes affect fetal growth, but recent work has begun to focus more on other contexts in which we find imprinted gene expression. For instance, imprinting has recently been demonstrated in flowering plants, in what appears to be a striking example of convergent evolution. In mammals, more research is addressing the effect of imprinted genes on cognition and behavior. Some of the chapters here give some indication of the puzzles that we will be faced with in these new frontiers of genomic imprinting research.

A Note about Terminology

The term "imprinted" has been used in the literature in different ways, often leading to situations in which different authors appear to be making contradictory statements. This is not an uncommon situation following the discovery of a novel phenomenon, as it often takes some time before a field agrees on a common set of terms and usage.

Some authors have used the phrase "imprinted allele" to refer to the allele that has been inactivated. Other authors have used "imprinted allele" to indicate an allele that has been epigenetically modified. If all parent-specific epigenetic modifications were transcriptional inactivators, there would be no ambiguity. However, at some loci, the addition of a particular epigenetic mark—such as DNA methylation—serves to activate expression of the allele, whereas the "unmodified" allele is transcriptionally silent.

Moreover, as our understanding of the regulation of imprinted genes has increased, it has become clear that in many cases it is meaningless to refer to either allele simply as "silenced" or "modified." Rather, most of these genes show extremely complex patterns of epigenetic modification and gene expression. Many imprinted genes occur in large clusters on the chromosomes. In these clusters, both parental copies typically receive some sort of epigenetic modification.

Exactly what terminology to use to discuss genomic imprinting has not been fully settled by the field. In this book, we have tried to consistently use the term "imprinted" to refer to a locus, rather than either of the alleles at the locus. An imprinted locus is one where alleles follow two different expression strategies conditional on parental origin. Alleles are referred to variously as "silenced," "methylated," "modified," etc.

The Structure of the Book

In assembling the set of topics and authors for this volume, I made the decision to favor depth over breadth. These chapters do not provide a comprehensive overview of all known imprinted genes. Part of the book is designed around the *Gnas* locus, which is representative of the mechanistic and phenotypic complexity that is likely to be associated with many imprinted genes. The rest of the book attempts to focus attention on some of the newer frontiers in genomic imprinting research.

The first chapter, by Lees-Murdock and Walsh, describes the current state of our knowledge of how the differential epigenetic marks associated with imprinted gene expression are first established in the male and female germ lines. This reprogramming occurs every generation and serves as the basis for the differential behavior of alleles throughout development and into adulthood.

The next four chapters focus particularly on the *Gnas* cluster of imprinted genes. Chapter 2, by Peters and Williamson, describes the complex set of molecular mechanisms that regulate imprinted gene expression in this region. Chapter 3, by Bastepe, describes the physiological and clinical effects of mutations associated with the various *Gnas* transcripts. Chapter 4, by Frontera et al, discusses the effects of imprinted genes on metabolism, particularly at the postnatal stage, where *Gnas* appears to play a central role.

Chapters 5 and 6 describe the cognitive and behavioral effects of imprinted genes. Chapter 5, by Davies et al, addresses these effects from a molecular-genetic perspective, particularly focusing on what we are learning by manipulating imprinted genes in the mouse. This chapter also provides our fourth and final perspective on the *Gnas* cluster, which encodes the neuron-specific Nesp transcript. Chapter 6, by Goos and Ragsdale, addresses the cognitive and behavioral effects of imprinted genes from a more clinical perspective, particularly how our understanding of these genes is informed by disorders in humans.

Chapter 7, by Garnier et al, describes what has been learned in the past few years about genomic imprinting in plants. Chapter 8, by Úbeda and Wilkins, takes a theoretical evolutionary perspective on the implications of imprinted gene expression for human disease. Finally, Chapter 9, by Mills and Moore, provides a summary and commentary on the ever-expanding collection of theories that have been proposed to explain the evolutionary origins of imprinting.

I would like to thank all of the authors for their hard work and patience on this project. I think that the result is a book that pushes at many of the important boundaries of our understanding of the phenomenon of genomic imprinting.

Jon F. Wilkins, Ph.D.

PARTICIPANTS

Murat Bastepe
Endocrine Unit
Department of Medicine
Massachusetts General Hospital
and
Harvard Medical School
Boston, Massachusetts
U.S.A.

William Davies
The Babraham Institute
Cambridge
U.K.
and
The Department of Psychological
 Medicine
University of Cardiff
Cardiff, Wales
U.K.

Benjamin Dickins
Laboratory of Developmental Genetics
 and Imprinting
The Babraham Institute
Cambridge
U.K.

Olivier Garnier
Genetics and Biotechnology Lab
Department of Biochemistry
Biosciences Institute
University College Cork
Cork
Ireland

Lisa M. Goos
Department of Psychiatry Research
The Hospital for Sick Children
The University of Toronto
Toronto, Ontario
Canada

Margalida Frontera
Laboratory of Developmental Genetics
 and Imprinting
The Babraham Institute
Cambridge
U.K.

Trevor Humby
School of Psychology
University of Cardiff
Cardiff, Wales
U.K.

Anthony R. Isles
The Department of Psychological
 Medicine
University of Cardiff
Cardiff, Wales
U.K.

Gavin Kelsey
Laboratory of Developmental Genetics
 and Imprinting
The Babraham Institute
Cambridge
U.K.

Sylvia Laoueillé-Duprat
Genetics and Biotechnology Lab
Department of Biochemistry
Biosciences Institute
University College Cork
Cork
Ireland

Diane J. Lees-Murdock
Stem Cell and Epigenetics Research
  Group
School of Biomedical Sciences
Centre for Molecular Bioscience
University of Ulster
Coleraine, Northern Ireland
U.K.

Walter Mills
Department of Biochemistry
Biosciences Institute
University College Cork
Cork
Ireland

Tom Moore
Department of Biochemistry
Biosciences Institute
University College Cork
Cork
Ireland

Jo Peters
MRC Mammalian Genetics Unit
Harwell, Oxfordshire
U.K.

Antonius Plagge
Physiological Laboratory
School of Biomedical Sciences
University of Liverpool
Liverpool
U.K.

Gillian Ragsdale
Leverhulme Centre for Human
  Evolutionary Studies
Cambridge
U.K.

Charles Spillane
Genetics and Biotechnology Lab
Department of Biochemistry
Biosciences Institute
University College Cork
Cork
Ireland

Francisco Úbeda
St. John's College
and
Oxford Centre for Gene Function
Oxford University
Oxford
U.K.

Colum P. Walsh
Stem Cell and Epigenetics Research
  Group
School of Biomedical Sciences
Centre for Molecular Bioscience
University of Ulster
Coleraine, Northern Ireland
U.K.

Jon F. Wilkins
Santa Fe Institute
Santa Fe, New Mexico
U.S.A.

Lawrence S. Wilkinson
Department of Psychological Medicine
and
School of Psychology
University of Cardiff
Cardiff, Wales
U.K.

Christine M. Williamson
MRC Mammalian Genetics Unit
Harwell, Oxfordshire
U.K.

CONTENTS

1. DNA METHYLATION REPROGRAMMING IN THE GERM LINE 1
Diane J. Lees-Murdock and Colum P. Walsh

Introduction 1
Methylation of Repeats and Genome Stability 3
Demethylation in the Germ Line 3
De Novo Methylation in the Developing Gametes 8
Fate of Methylation Differences Inherited from the Gametes in the Early Embryo ... 10
Methylation Enforces Transcriptional Silencing and Suppresses Recombination 11
Repeats Attract Methylation, while Transcription Factors Block It 12
Conclusions 12

2. CONTROL OF IMPRINTING AT THE *GNAS* CLUSTER 16
Jo Peters and Christine M. Williamson

Background 16
Extent of the Region 17
Transcripts at the *Gnas* Locus 18
Proteins Encoded at the *Gnas* Complex Locus 18
Function of Proteins Encoded at the *Gnas* Locus 18
Imprinting Centers 19
The *Exon 1A* DMR Controls the Expression of *Gnas* 21
The *Nespas* DMR is the Principal ICR in the *Gnas* Cluster 21
Interaction between the *Nespas* DMR and the *Exon 1A* DMR 22
The Role of *Nespas* 23
The Role of the *Exon 1A* DMR 23
Conclusions 24

3. THE *GNAS* LOCUS AND PSEUDOHYPOPARATHYROIDISM 27
Murat Bastepe

Introduction 27
Inactivating Gsα Mutations and Multiple Hormone Resistance: PHP-Ia 29
Role of Tissue- and Parental Origin-Specific Gsα Expression
  in Hormone Resistance 30
Mutations Affecting the Imprinting Control Regions of *GNAS*
  and PTH-Resistance: PHP-Ib 32
Conclusion 36

4. IMPRINTED GENES, POSTNATAL ADAPTATIONS AND ENDURING EFFECTS ON ENERGY HOMEOSTASIS 41

Margalida Frontera, Benjamin Dickins, Antonius Plagge and Gavin Kelsey

Introduction 41
Imprinted Gene Syndromes and Obesity 41
Genetic Evidence for Parent-of-Origin Effects on Obesity 43
Imprinted Gene Action in the Hypothalamus 43
Imprinted Gene Action in Adipose Tissues 44
The *Gnas* Locus 46
The Role of Gsα in Energy Homeostasis 48
XLαs in Postnatal Adaptations and Metabolism 53
Mutations of the *GNAS* Locus in Human Neonatal Physiology
   and Adult Energy Homeostasis 54
The 'Conflict Hypothesis' and Beyond 55
Concluding Remarks 56

5. WHAT ARE IMPRINTED GENES DOING IN THE BRAIN? 62

William Davies, Anthony R. Isles, Trevor Humby and Lawrence S. Wilkinson

Abstract 62
Imprinted Genes and the Brain 62
Summary Evidence for a Role for Imprinted Genes in Brain Function 63
Characteristics of Brain-Expressed Imprinted Genes 63
Imprinted Gene Effects on Brain Development 64
Imprinted Gene Effects on Behavior 65
Through What Mechanisms Might Imprinted Genes Affect (Adult) Behavior? 66
Imprinted Genes in the Adult Brain 67
What Adult Behaviors Will Imprinted Genes Influence? 67

6. GENOMIC IMPRINTING AND HUMAN PSYCHOLOGY: COGNITION, BEHAVIOR AND PATHOLOGY 71

Lisa M. Goos and Gillian Ragsdale

Abstract 71
Genomic Imprinting in Human Cognition and Behavior 71
Imprinted Syndromes, Behavioral Phenotypes and Neuropsychological Research 76
Conclusion 80

7. GENOMIC IMPRINTING IN PLANTS 89

Olivier Garnier, Sylvia Laouiellé-Duprat and Charles Spillane

What is Genomic Imprinting? 89
Evolution of Genomic Imprinting 89
Genomic Imprinting in Plants 90
Imprinting Regulation at the Maternally Expressed *MEDEA* Locus
   in *Arabidopsis thaliana* 91

Imprinting Regulation at the Maternally Expressed *FWA* Locus
  in *Arabidopsis thaliana* 93
Imprinting Regulation at the Paternally Expressed *PHE1* Locus
  in *Arabidopsis thaliana* 93
Imprinting Regulation at the Maternally Expressed *FIS2* Locus
  in *Arabidopsis thaliana* 93
Differentially Methylated Domains (DMDs) and Imprinting Regulation in Plants 93
Emerging Models for Imprinting Regulation in Plants 94

8. IMPRINTED GENES AND HUMAN DISEASE: AN EVOLUTIONARY PERSPECTIVE 101

Francisco Úbeda and Jon F. Wilkins

Abstract 101
Do Disorders Linked to Imprinted Genes Share a Common Motif? 102
Growth and Resource Acquisition 103
Post-Natal Behavior 107
Cancer 108
Are Imprinted Genes Particularly Fragile? 108
Mutations 109
Epimutations 111
Uniparental Disomies 112
Implications for the Prevention and Treatment of Human Disease 112

9. EVOLUTIONARY THEORIES OF IMPRINTING— ENOUGH ALREADY! 116

Tom Moore and Walter Mills

Abstract 116
Introduction 116
What Needs to be Explained? 117
The Etiquette of Proposing a New Theory of Imprinting 119
Conclusion 121

INDEX 123

CHAPTER 1

DNA Methylation Reprogramming in the Germ Line

Diane J. Lees-Murdock* and Colum P. Walsh

Abstract

In mammals, methylation occurs almost exclusively on the CpG dinucleotide in DNA and shows no preference for sequence context surrounding this target. CpGs are found on many different sequence classes and methylation of this dinucleotide is associated with repression of transcription. Reprogramming methylation in the primordial germ cells establishes monoallelic expression of imprinted genes which exhibit monoallelic expression throughout the lifetime of an organism, maintains retrotransposons in an inactive state and inactivates one of the two X chromosomes. In addition to direct transcriptional silencing, DNA methylation is important for suppression of recombination, and resetting this information is therefore necessary for maintenance of genomic stability. In this chapter, we will review the recent progress in our understanding of the time course and extent of DNA methylation reprogramming of many different sequence classes. We focus on the mouse germline, since this has been the model system from which we have gained the most knowledge of the process. In addition we will examine some of the evidence suggesting a link between repeat methylation and methylation of epigenetically controlled single-copy genes. To do this, we will look at the temporal sequence of methylation events from the time the germ cells become recognizable as a discrete population until the mature male and female gametes fuse and form the early embryo.

Introduction

Diverse types of repetitive DNA elements and epigenetically-controlled genes (such as the imprinted genes and genes on the inactive X chromosome in mammals) undergo DNA methylation reprogramming in the mouse germ line. CpGs are distributed differently in the different sequence classes, with single-copy sequences being generally deficient in CpGs except for a cluster of sites known as a CpG island, which can often be found at the promoter region (see Table 1). Most CpG-island genes, such as *Pax3* or *Oxtr*, escape methylation in all tissues and thus are not influenced by changes in methylation or in methyltransferase level,[1] except in the exceptional cases where such a gene may become aberrantly methylated and silenced in cancer cells. The best example of this is *MLH1* in human sporadic colon cancer.[2]

Self-replicating interspersed repeats such as Intracisternal A particle (IAP) and LINE1 (L1) elements, on the other hand, often have relatively high levels of CpGs throughout, including in the transcriptional control regions. For imprinted genes, an additional CpG island—often at some distance from the promoter—known as the Imprint Control Region (ICR) is crucial for regulating the gene. The ICR bears the imprint, with one copy being methylated in a parent of origin-specific manner. The absolute proof that an ICR controls imprinting requires functional disruption by

*Corresponding Author: Diane J. Lees-Murdock—Stem Cells and Epigenetics Research Group, School of Biomedical Sciences, Centre for Molecular Bioscience, University of Ulster, Coleraine, N. Ireland, BT52 1SA U.K. Email: dj.lees@ulster.ac.uk

Genomic Imprinting, edited by Jon F. Wilkins. ©2008 Landes Bioscience and Springer Science+Business Media.

Table 1. Sequence classes affected by changes in DNA methyltransferase levels

Sequence Cswlass	Sequence Type	CpG Location	Effect of Methylation	Examples	Selected References
Single copy genes	Imprinted genes	a) Promoter b) Imprint Control Region	Transcriptional silencing Enhancer blocking/insulation	H19 H19/Igf2 ICR	72 73,74
	X chromosome genes	Promoter	Transcriptional silencing	Hprt	75
	Autosomal, non-imprinted	Promoter	Transcriptional silencing in some cancer cells	MLH1	2
Repeat sequences	Interspersed repeats	Many and interspersed	a) Transcriptional silencing at promoter/enhancer b) Increased stability (decreased recombination?)	IAP, L1	12,13,51 12,62
	Satellite repeats	Many and interspersed	Increased stability	Sat2	4-6
	Subtelomeric repeats	Many and interspersed	Increased stability, decreased recombination	Chr. 19 repeats	7
	Microsatellite repeats	None	Increased stability	CA_{17}, A_{25} repeats	8-10

homologous recombination or other means. In the absence of such definitive data putative ICRs are sometimes called Differentially Methylated Regions (DMRs).

Methylation of Repeats and Genome Stability

The DMRs of imprinted genes are CpG islands, but on the whole the single-copy regions of the genome are CpG-poor. Many repeat sequences on the other hand are CpG-rich (see Table 1) and contain the bulk of the total CpG content of the mouse genome.[3] Therefore there has been a good deal of interest in the methylation dynamics of these sequences during gametogenesis as well.

Importantly, it is becoming increasingly clear that DNA methylation in mammals is associated with repeat stability: demethylation of minor satellites, subtelomeric satellites, microsatellites and selfish repeats appears to lead to increased recombination and may result in destabilisation of the chromosome on which they reside (Table 1 and refs. therein). The importance of maintaining methylation on various types of repeats has been demonstrated in a number of methyltransferase-deficient systems (see also Table 1).

Immunodeficiency, centromeric instability and facial anomalies (ICF) syndrome in humans for instance is caused by mutations in the *Dnmt3b* gene.[4-6] Patients with ICF syndrome display hypomethylation of classical satellites and some interspersed repeats and this results in chromosomal instability, with a range of effects including the generation of multiradial chromosomes and duplications or deletions of whole chromosome arms. The chromosomes affected are those that contain sat2 or sat3 sequences in the pericentromeric regions and the aberrant copies show expansion and recombination between the repeat regions.[4]

Cells deficient in methyltransferase activity also show increased rates of recombination at telomeric sequences,[7] despite the fact that the mammalian telomeric sequence TTAGGG does not contain the target CpG. However, subtelomeric sequences are CpG-rich and normally methylated, but show hypomethylation in the methyltransferase-deficient cells,[7] so this may play a role in controlling telomere length.

Another repeat class that has been shown by three different groups to show decreased stability in DNMT1-deficient cells are microsatellite repeats.[8-10] These simple, interspersed repeats usually consist of runs of 1-3 nucleotides which lack a target CpG. Interestingly, in Dnmt1 knockout cells microsatellites of various types, many having no CpG target, show decreased stability. The mechanism involved here is currently unclear.

Mobilization of transposons may also facilitate recombination between nonhomologous loci[11] leading to deletions and translocations. In Dnmt3L mutant males, meiotic instability and aberrant, branching synaptonemal complexes were seen in combination with demethylation and increased transcription of IAPs, L1s and a de novo ERV1 LTR insertion.[12,13]

Suppression of recombination by DNA methylation has been demonstrated in plants and fungi (see ref. 14) and it seems likely that methylation may play a similar role in mammals in addition to its more familiar role in direct transcriptional silencing.

Demethylation in the Germ Line

DNA methylation is heritable through cell division and can even be passed from one generation to the next via the egg and sperm (see also the chapter by Ubeda and Wilkins). However, methylation at imprinted loci, which is often crucial for normal embryonic growth, must be reset during germ cell development, since the early diploid cells, known as primordial germ cells (PGCs), will develop into the haploid egg or sperm according to the sex of the newly developing organism they are located in. Likewise, reprogramming of methylation on the inactive X and presumably on any autosomal genes that have become methylated must also occur. The various classes of repeat sequences are fully methylated in adult somatic tissues and the results summarised in the section above show that many repeat sequences destabilize with sometimes catastrophic effects in the absence of DNA methylation. The genome must then be protected from hypomethylation of repeats as much as possible during periods of reprogramming.

Methylation Status of Migrating Germ Cells

The primordial germ cells arise as a small cluster of alkaline phophatase-positive cells underneath the allantois in the post-implantation mouse embryonic epiblast at embryonic day 7.5 (e7.5).[15] These precursor cells migrate to the developing gonads where they arrive as mitotically dividing cells around midgestation e10.5. Differentially methylated regions (DMR) of imprinted genes in e10.5 PGCs (prior to sexual dimorphism) retain the methylation patterns inherited from the egg and sperm (see Fig. 1A,B) and interspersed repeat sequences (see Fig. 1C,D) exhibit methylation levels similar to adult somatic cells.[16-18] In females, only one X chromosome is active at the migratory PGC stage,[19] the other one having already undergone X-inactivation in the epiblast to provide dosage compensation with respect to the male, which has only one X and the gene-poor Y chromosome which determines maleness. Although other repeat classes such as minor satellites have not been examined, the evidence from other repeats indicates that methylation levels on the whole in germ cells are similar to those in their somatic neighbours (see Fig. 1C).

Demethylation of Imprinted Genes

Early studies of methylation status of imprinted genes in primordial germ cells were hampered by the need to maintain these cells in culture during the isolation procedure. This process affected the methylation levels of imprinted genes, and culture conditions are now known to be crucial for maintaining correct imprinting in isolated PGCs and in Assisted-Reproductive Technology (ART) embryos.[20-23] As a result it took some time to obtain a clear picture of methylation events in the germ line.

It is now clear from studies carried out on PGCs isolated without culture—and directly analysed by bisulfite sequencing following isolation—that the DMRs of imprinted genes, including the maternally methylated *Snrpn* DMR1, *Peg3*, *Lit1* and *Igf2* and the paternally methylated *H19* and *Rasgrf1* are synchronously demethylated between e11.5 and e12.5.[16-18] Non-imprinted single-copy sequences also become demethylated at this point in development (Fig. 1F).[16] Interestingly, for the paternally methylated *Gtl2* gene, complete erasure is not achieved at e12.5.[17] It is maintained in a partially methylated state until it becomes fully methylated again in the male germ line. It continues to lose methylation slowly in the female germ line until the fully unmethylated pattern seen in the mature oocyte is achieved (Fig. 1B).[17,24,25] The dynamics of methylation of this gene are more reminiscent of the repeat sequences, which we will deal with now.

Partial Removal of Methylation on Repeat Sequences

Three different classes of repeats have been closely examined. Two of these are interspersed selfish DNA elements capable of transcription and retrotransposition. IAPs are endogenous retroviral elements containing long terminal repeats (LTRs) with all the components necessary for replication and integration of the element into a new genomic site. L1 elements contain a 5' promoter and two open reading frames, but LTRs are absent.[26] Both of these types of repeat element are found dispersed throughout the genome. Recent studies have focused on the LTR of the IAP and the 5' promoter of the L1, where methylation is known to affect transcription. These studies have focused particularly on the subclasses of the elements thought to have been most recently active (see ref. 21). The third type of repeat studied in the germ line is the nontranscribed minor satellite sequence, which is found close to the centromere in 20-200bp repeats.[27] The other repeats known to be affected in DNA methyltransferase mutants have not yet been studied in any detail in gametogenesis and so are not dealt with here.

The repetitive DNA elements studied so far are co-ordinately demethylated in the male and female genomes upon entry into the gonad. However, in contrast to the imprinted sequences, which are completely demethylated, the repeat sequences only lose some methylation, retaining a substantial degree, with the minor satellites (Fig. 1C) retaining the most and L1 sequences (Fig. 1D) the least during all stages of development examined.[21] In the male germline IAPs (Fig. 1C), L1s and minor satellites are partially demethylated in parallel with the imprinted genes and retain their undermethylated state until e15.5 when the male PGCs, now known as prospermatogonia, have entered mitotic arrest and de novo methylation of all these sequences will occur (see *Partial removal of methylation on repeat sequences*).[16,21]

In the female germline demethylation continues past e15.5 as the female PGCs, now known as primary oocytes, progress through prophase of meiosis.[21] At least one study using bisulfite sequencing shows even lower levels of methylation of IAPs in nongrowing oocytes isolated from mice one day after birth,[28] but these sequences will undergo methylation during oocyte growth (see *Partial removal of methylation on repeat sequences* and Fig. 1C). L1 elements lose methylation most rapidly, with more than half the elements showing demethylation by e17.5,[21] and these sequences appear to remain at this level of methylation in the mature oocytes (Fig. 1D).[29,30]

X-Reactivation

The epigenetic marks on the X chromosome of female mammals also undergo dynamic reprogramming during germ cell development. Early studies of the timing of X-inactivation in mice took advantage of the fact that the inactive X chromosome forms a distinct heterochromatin body in the nucleus (the Barr body), and later work was facilitated by the use of transgenes inserted on the X chromosome encoding marker proteins such as X-gal.[19,31-33] These studies found that one X was inactive in migrating PGCs, but becomes reactivated on arrival at the gonadal ridge, presumably reflecting both demethylation and alteration of any other epigenetic marks that may be associated with inactivation, such as histone changes, although no methylation analyses of the transgenes in germ cells were done. This reprogramming would be consistent with a need to reprogram the X for the next generation so that two inactive or two active copies are not inherited, in the same fashion as for the imprinted genes.

Mechanistically, this process seems to be controlled by the action of the genes *Xist* and *Tsix*. These share many features of imprinted genes, including a paired antisense/sense arrangement, presence of a DMR containing CTCF-binding sites and parent-of-origin specific expression in some tissues (see *Methylation of the inactive X*). The *Xist* gene is actively transcribed only on the inactive X (Xi-specific transcript) with its RNA coating the inactive chromosome. This seems to form part of the inactivation mechanism. *Tsix* is transcribed on the active X and prevents *Xist* transcription on that chromosome. Methylation of the recently-identified DMR on the inactive X prevents *Tsix* expression and allows *Xist* expression, in a way that is similar to the *H19/Igf2* pair (see 34 and refs. therein). Both copies of *Xist* are silent in cells derived from embryos of *Dnmt1* knockout mice. In these knockouts, the control region is demethylated, and both X chromosomes become active. Presumably *Xist* is demethylated in e12.5 PGCs and the inactive X becomes re-activated at this stage. Previous studies[16] have not examined the newly-characterised DMR.[34]

The Demethylation Debate

The observation that most of the erasure of imprints and methylation of single copy sequences seems to occur in a very short space of time, perhaps as little as one cell division, while the maintenance methyltransferase Dnmt1 is still present in the nucleus has led to speculation that demethylation is an active process in the germ cells.[16] However, biochemical data suggests that an enzymatic reaction catalysing the direct reversal of the methylation reaction is energetically unfavourable.[35,36] There are a number of alternative mechanisms which may be possible (see ref. 37). In flowering plants, active demethylation in the germ cells has been clearly demonstrated, but utilises a glycosylase component of the base excision repair pathway called DEMETER (DME).[38] In mouse PGCs and oocytes cytosine deaminase expression has been observed,[39] but no evidence has so far been produced to suggest that any of the above type of enzyme activities affects imprinted DMRs in mouse germ cells.

Recent data examining the allele-specific methylation status of the imprinted genes *H19* and *Snrpn* in both migratory and post-migratory PGCs (e9.5-e11.5) indicates that the erasure process may take longer than previously thought.[18] Before arrival at the gonad (e9.5), most of the paternally inherited alleles are hypermethylated at the *H19* DMR, but a small percentage of these are hypomethylated (19% of paternal alleles with less than 50% methylation).[18] Hypomethylation of the paternal *H19* DMRs increases incrementally until e11.5 when over 75% of paternal alleles are

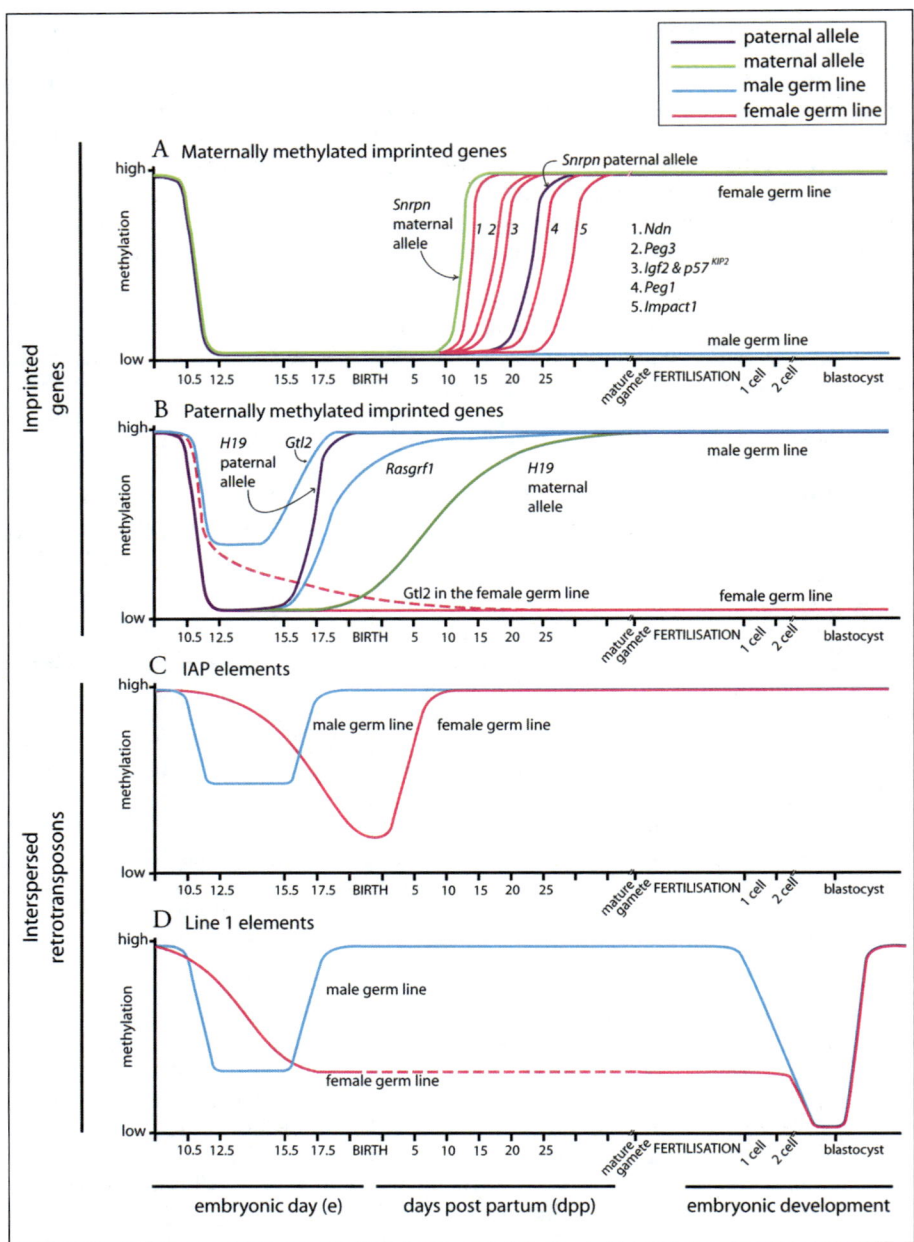

Figure 1. Please see legend on following page.

hypomethylated. These results suggest three possible mechanisms for erasure of imprints. The first is that upon entry into the gonad, the DMR is rapidly demethylated via an active process occurring within one cell division (Fig. 2A). The slow increments of demethylation observed from e9.5 onwards could be explained if PGCs colonize the gonad gradually, with their migration continuing through e11.5.[40] Alternatively erasure of imprints may take place over the equivalent of four

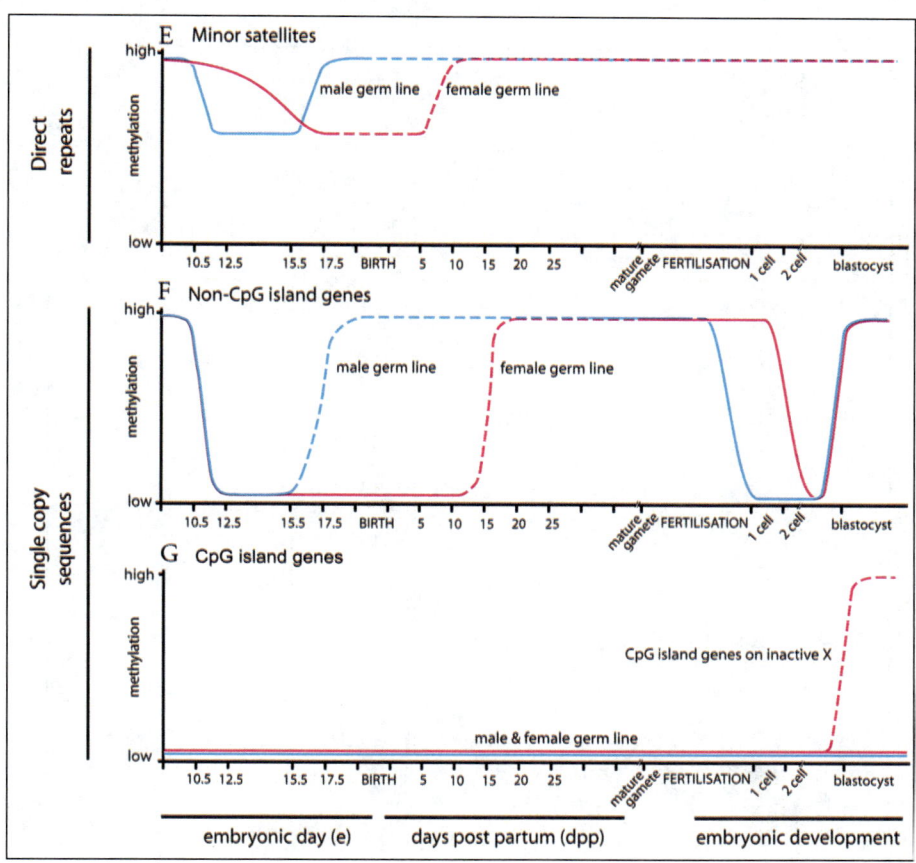

Figure 1 (A-D viewed on previous page). Methylation dynamics of various sequence classes in the mouse germ line and early embryo. All sequences undergo demethylation in the post-migratory germ cells at embryonic day 12.5 (e12.5). De novo methylation occurs at different times in the two germ lines: it begins at e15.5 and is complete around birth in the male germ cells before the appearance of the Type A spermatogonial stem cell population. In the female, de novo methylation does not occur until the primary oocytes, which are paused in the diplotene (dictyate) stage of meiosis, are recruited to grow: the first and largest cohort enter growth around ten days after birth and methylation dynamics are indicated for this group. After fertilisation, demethylation is seen in the early embryo for some sequences, followed by a second embryonic wave of de novo methylation after implantation. For imprinted genes, differences between the maternal and paternal alleles in the timing of methylation are indicated where known. Dashed lines indicate where data is incomplete. The blue line represents methylation levels in the male germ line before fertilisation and the paternal genome during early embryonic development. Methylation levels in the female germ line and maternal genome are represented by the pink line. Purple and green represent the paternal and maternal alleles (respectively) of imprinted genes.

to five PGC doubling times via a passive process[19] initiated either before or after colonization of the genital ridge (Fig. 2B,C). Migrating PGCs from transgenic mice expressing green fluorescent protein specifically in the germ line also begin demethylation of the *Igf2r* DMR2, albeit slowly, at e9.5. Following colonization of the genital ridge, demethylation is rapid.[41]

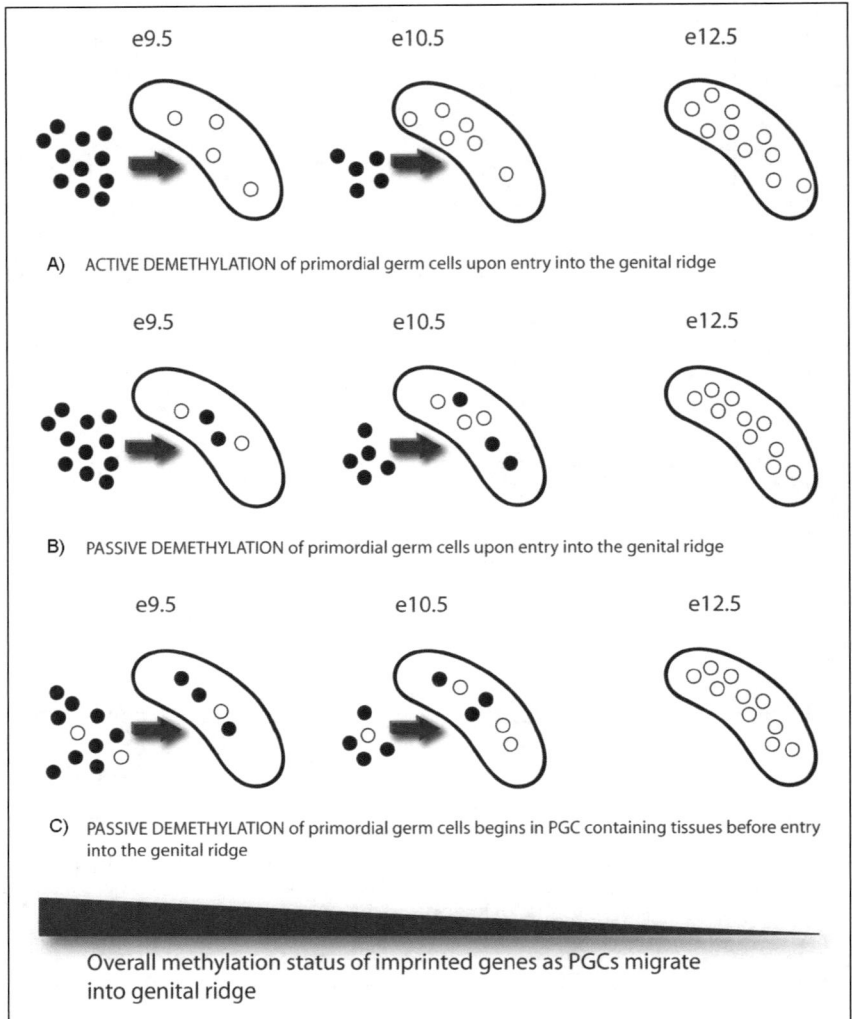

Figure 2. Possible mechanisms for demethylation of imprinted genes in primordial germ cells. Black circles represent PGCs with methylation present at the DMR of imprinted genes, white circles represent PGCs with unmethylated DMR.

The mechanism of erasure in mammals has not yet been confirmed and no reproducible demethylation activity has been identified, therefore the debate over the mechanism of imprint erasure will continue until demonstration of an enzymatic reaction that can catalyze demethylation in animals.

De Novo Methylation in the Developing Gametes

Following erasure of imprints and partial erasure of methylation on repeats, these sequences must undergo de novo methylation during subsequent germ cell development to achieve the methylation patterns observed in the mature gametes.

Methylation of Imprinted Genes in the Male Germline

There are a very limited number of imprinted genes that exhibit methylation in the male germ line,[42] the best-studied being *H19*. Following erasure of methylation, the germ cells are still diploid and contain two copies of each gene, each of which needs to be methylated to ensure that each haploid sperm is correctly imprinted. Although both copies of *H19* have had their previous methylation pattern erased, it appears that they still retain another epigenetic mark, since the paternally-inherited *H19* becomes methylated faster than the maternal allele.[43-45] De novo methylation is initiated at e14.5 on the paternal allele only and completed by e15.5.[44] The maternal DMR becomes does not acquire full methylation until around birth.[44-46] Methylation appears to be complete in the post-natal gonocyte before meiosis starts. Both *Rasgrf1* and *Gtl2* acquire methylation following a timecourse similar to *H19* (see Fig. 1B).[17]

Methylation of Imprinted Genes in the Female Germline

While in the male germ line resetting of methylation occurs before meiosis, maternal ICRs are hypomethylated until after the pachytene stage of meiosis I, which occurs in the postnatal growing oocyte (Fig. 1A).

An early indication that maternal methylation was established during this stage came from studies using MII oocytes containing transplanted nuclei of oocytes from different growth stages. These oocytes were fertilized and cultured in vitro until the blastocyst stage, when they were transferred to recipients to assess developmental potential.[47] Donor nuclei from oocytes in the latter half of growth were able to support post-implantation development while early-growth-phase oocyte nuclei were not. Parthenogenetic embryos were also created from the genomes of oocytes at different stages of growth. The methylation status of several imprinted genes in the embryos was analysed and the methylation of each imprinted gene examined was found to occur at a slightly different stage of oocyte growth.[48]

The first study examining the methylation state of oocytes directly isolated from the ovary using the sensitive bisulfite sequencing technique confirmed that the oocyte growth phase corresponds to the period of maternal imprint establishment on the *Snrpn* gene.[49] Two further detailed studies tracking methylation using F1 mice, where the maternal and paternal alleles could be distinguished by single nucleotide polymorphisms, extended these findings to a larger set of genes, including *Igf2r*, *Peg1*, *Peg3* and *Zac1*, which are all located in different chromosomal regions. The methylation level is correlated to the size of the oocyte, being gradually acquired from the onset of growth. By the time the oocytes reach 55-60μm the DMR has become heavily methylated.[28,50] For three genes, *Snrpn*, *Zac1* and *Peg1*, the maternally inherited allele appears to acquire methylation before the paternal allele, with the evidence for this being most robust for *Snrpn*. This is another indication (see above) that a separate epigenetic mark may be retained in the absence of DNA methylation at the DMR that still allows the parent of origin of each allele to be distinguished.

Remethylation of Repeat Sequences

In the male germ line diverse classes of repeat sequences undergo coordinated de novo methylation in a brief window from e15.5 to e17.5, ensuring that IAPs, L1s and minor satellite selfish DNA sequences are inherited from the sperm in a fully methylated state.[21,30,51] In contrast, there is no de novo methylation event at e17.5 in the female germ line and the repeats remain hypomethylated at this stage.[21]

IAP sequences become methylated again prior to the maternal *Snrpn* allele in the growing oocytes and are fully methylated in the mature MII oocyte (see Fig. 1C).[30,51] Line1 elements may escape de novo methylation during oogenesis as they appear to be inherited from the oocyte in a hypomethylated state (see Fig. 1D).[29,30] The epigenetic status of minor satellites in the mature oocyte has not been examined and the timing of remethylation of these sequences has not been defined but is likely to occur in the growing oocyte (see Fig. 1C).

Developmental regulation of the DNA methyltransferase enzymes (see ref. 52) ensures that nuclear expression of Dnmt3a and Dnmt3b proteins are concomitant with high levels of

Dnmt3L transcription specifically during these separate periods of de novo methylation in each germ line.[53]

Methylation of the Inactive X

X inactivation in the preimplantation embryo is imprinted and methylation is thought to play a role in this phenomenon (see ref. 54 for a review). The paternal X is preferentially inactivated while the maternal X is active. Following demethylation events in the zygote (see *Fate of methylation differences inherited from the gametes in the early embryo*), X inactivation is then random in embryonic tissues but remains imprinted in the extraembryonic lineages.

The DMR at *Xist/Tsix* contains CTCF binding sites and is inherited from the sperm hypermethylated and from the oocyte in a hypomethylated state.[34] The question of when methylation is established on the X chromosome in the male germ line is an interesting one and, surprisingly, largely unexamined. This event presumably occurs during the period of de novo methylation between e15.5 and birth, since it is inherited in a hypermethylated state in sperm.[34] However, this might be thought to trigger inactivation of the X chromosome present in the Type A spermatogonial stem cells, the absence of which is known to be lethal to cells, so inactivation is unlikely to occur at this stage. Likewise, methylation of single-copy CpG-island genes on the X during this period would silence them, with the same effect. Methylation of selfish repeats would on the other hand be necessary on this chromosome to inactivate them and prevent de novo insertions in the germ line. To our knowledge, no studies have examined the methylation status of any of these sequences on the X during male germ cell development. In the female germ line, both X chromosomes are reactivated in the e12.5 primary oocytes, but since the cells enter meiosis immediately afterwards, transcription would be shut off anyway until the dictyate stage of prophase I, when the oocyte becomes very transcriptionally active (this is the stage in which lampbrush chromosomes are seen in frog oocytes). Double doses of X are also lethal, however recent evidence suggests that the diploid primary oocytes cope by down-regulating each X to half its normal transcription level.[55] Methylation of L1 and IAP elements on the X in oocytes presumably follows the dynamics seen for their counterparts on the autosomes (see *Remethylation of repeat sequences*), although again this has not been examined in any detail.

Fate of Methylation Differences Inherited from the Gametes in the Early Embryo

Early studies of DNA methylation changes in the early embryo suggested a wave of demethylation in the preimplantation embryo (see ref. 3). These studies and those utilising methylated cytosine antibodies, indicate that bulk DNA is losing methylation until implantation. There is some evidence to suggest that the paternal genome undergoes active demethylation in the one cell embryo, while the maternal genome loses methylation passively at each cell division until the blastocyst stage.[30,56,57] This also seems to be the case for the L1 sequences and presumably for single-copy, nonCpG-island genes like *Acta1* and *Prf1*, though they have not been examined in detail in this period. Following implantation, these sequences all start to become methylated in the wave of de novo activity concomitant with *Dnmt3a* and *Dnmt3b* expression from the egg cylinder stage on. For the single-copy genes, methylation is unlikely to be complete for many loci until after birth in the mature tissues, since neonatal mice show no methylation at most sites examined in non-expressing tissues. Here methylation follows transcriptional silencing and most likely serves to stably lock in the silent state.[1]

If DNA methylation is to act as an imprint, however, it is essential that after fertilization, methylation patterns established on imprinted genes in the gametes must escape the early wave of demethylation in order for the methylation to act as a signal for parent of origin when the genes become transcriptionally active. For those genes which have been followed in detail (still surprisingly few) this does appear to be the case.[42,58] Tellingly, in the absence of Dnmt1 activity in the early embryo, methylation acquired in either the male (e.g., *H19*) or female (e.g., *Igfr2*) germ line can be lost during cleavage stages with consequent loss of imprinting and death of the embryo.[59]

IAP elements, like the imprinted genes, escape demethylation in the zygote, whereas L1 elements do not.[29,30] Methylation of the satellite sequences may occur either in the growing oocyte at the same time as the IAP elements or after the blastocyst stage, concurrently with the L1 elements. Either way, these repeats are fully methylated in the e9.5 embryo.[6]

Thus by the time the germ cells of the new embryo arise from the base of the allantois at the egg cylinder stage, the methylation pattern seen in adult tissues has been established once more on all the sequence classes, before the cycle of demethylation and remethylation in the germ cells begins again.

Methylation Enforces Transcriptional Silencing and Suppresses Recombination

Do the changes in methylation seen in the germ cells make sense given what we know or suspect about DNA methylation? On the whole, the answer seems to be yes.

The ICRs associated with imprinted genes are demethylated in post-migratory germ cells, then methylated de novo in the maturing gametes to reset the imprint. This ensures correct dosage compensation in the next generation, which will inherit one active and one inactive allele. The methylation mark distinguishes the inactive and active alleles and is untouched by the wave of demethylation and remethylation in early embryo, as would be expected.

Like the autosomal imprinted genes, the inactive X is reactivated, then methylated in the male germ line at *Xist/Tsix* to ensure inactivation of the paternal X in the extra-embryonic tissues of any subsequent female embryos and correct dosage compensation. However, the methylation difference between the paternal and maternal X is not immune to the reprogramming that occurs in the early embryo.[34] This explains why after implantation no parent-of-origin effect is seen and random X inactivation occurs, with methylation of the CpG island-containing genes such as *Hprt* on the inactive X driven by the embryonic methylation machinery.

Non-imprinted autosomal genes that have a CpG island at the promoter are unmethylated at all stages and so unaffected by methylation changes. Data is incomplete on those autosomal genes with no CpG island, but indications are that their transcription is independent of methylation and that they become passively methylated and demethylated along with the repeats that make up the rest of the genome.[1]

The selfish DNA repeats interspersed through the genome, such as the IAPs and L1s, are generally kept in a highly methylated state in most tissues, consistent with the fact that demethylation of these sequences can allow transcriptional reactivation.[12,51] In the male germ line, which forms a stable stem cell population in the adult, partial demethylation is seen for a brief window between e12.5 and e15.5. By e17.5, when the mature stem cells begin to appear, the sequences are fully methylated again. If this is prevented, by knocking out Dnmt3L, L1s, IAPs and the ERV endogenous retrovirus become active and can cause de novo insertions.[12,13] Female germ cells enter meiosis immediately after demethylation and are transcriptionally inactive. Demethylation appears to continue as they progress through prophase I of meiosis up to the dictyate stage for IAPs, at which point the oocytes can remain in meiotic arrest until recruited to grow and mature before ovulation. IAPs are methylated early during oocyte growth,[28] thus preventing widespread transcription. For L1s, demethylation may persist on the maternal chromosomes until post-implantation stages,[29,30] and it would therefore appear that some other mechanism must prevent transcription of these elements during oocyte growth.

This latter is another piece of evidence suggesting that epigenetic marks other than methylation participate in suppressing transcription. We saw in earlier sections that maternally and paternally derived alleles of the imprinted genes can still be distinguished after the removal of DNA methylation at the ICR, since they become de novo methylated at different rates. Likewise, parent-of-origin specific X-inactivation occurs in marsupials in the absence of marked DNA methylation differences,[60] and recent evidence from methyltransferase mutants suggest that post-implantation inactivation in eutherians is also initiated by a mechanism other than methylation.[61] Methylation can be seen rather as a stabilizing mechanism locking in transcriptional silencing.[31]

Regions of the genome that are transcriptionally silent are likely to be packaged into inert chromatin with concomitant suppression of recombination, which is often coupled to transcription. In line with this, demethylation of DNA brought about by different methyltransferase knockouts can lead to the decondensation, breakage and rejoining of pericentric repeat arrays,[4-6] and an increase in recombination rates at telomeric repeats.[7] Similarly, decreases in DNA methylation have been shown to increase mitotic recombination at de novo transgene insertions at multiple locations in the genome,[62] and lead to increased nonhomologous recombination during male meiosis.[12]

Repeats Attract Methylation, while Transcription Factors Block It

DNA methyltransferases in mammals have little or no intrinsic sequence preference,[63] and appear to methylate CpG when the target is available to them. This is borne out by the fact that methylation of almost all the diverse sequence classes occurs simultaneously in the germ cells and post-implantation embryo, suggesting that sequence context is not an important factor. Evidence for this is also clear in the human DNA sequence, where CpG has been severely depleted in almost all contexts as a result of deamination of methylated CpG (see ref. 37 for a review). The exceptions are CpG islands, which must therefore be protected from methylation in germ cells (see Fig. 1G). CpG islands are often found in the promoter regions of genes involved in housekeeping or early development. These promoters are thought to normally be occupied by a basal transcription complex assembly, which would most likely protect the CpGs from methylation. The only cases where CpG islands are methylated are at imprinted genes, retrotransposons and on the inactive X.

Methylation or occupancy of the promoter by transcription factors can be seen as mutually exclusive. If methylation occurs first, then the transcription factors cannot bind, but if a factor is bound then methylation cannot occur. This is well illustrated by the case of CTCF at the *H19/Igf2* ICR. Depletion of this factor from the growing oocyte allows de novo methylation and transcriptional silencing of the maternal copy of *H19*.[64] This model would also predict that in CTCF-depleted oocytes, methylation of the *Xist/Tsix* CTCF sites will occur and prevent imprinted X-inactivation in extra-embryonic tissues of the early embryo.[34] CTCF sites at the *Rasgrf1* ICR function in a similar fashion.[65]

On the inactive methylated allele of *Rasgrf,1* the presence of repeats attracts methylation. If these are removed imprinting becomes faulty, but they can be functionally replaced by repeats from the *Igf2r* ICR.[66] The repeats from the *Igf2r* ICR can also confer imprinting on several transgenic lines.[67] It has recently been suggested by Mary Lyon that repeat elements, in particular L1s, present on the X chromosome may be important for the spreading of the inactivation signal by attracting methylation and/or other epigenetic marks.[68,69] Some recent papers have lent weight to this theory.[61,70,71]

However, methylation appears to be a stochastic, rather than a processive, event, suggesting that the density of CpGs may be more important than any particular repeat structure. The repeats close to the *Rasgrf1*, *H19* and *Gtl2/Dlk* ICR are all quite different,[17] and the dynamics of their demethylation and remethylation differ significantly (see Fig. 1B), but all eventually do become methylated. Rather than looking for sequence motifs that attract methylation, we should instead be looking for motifs, like CTCF binding sites, that appear to block it.

Conclusions

It is important that DNA methylation profiles are correctly established and maintained on all classes of sequence found in the genome. In addition to establishment of imprints that are an absolute requirement for development to proceed, methylation must be retained as much as possible for repeat sequences, to prevent transcriptional derepression of parasitic elements and deregulated recombination at repeat sequences, with disastrous consequences for the genome.

Acknowledgements

The authors wish to thank members of the Stem Cells and Epigenetics Research Group at the University of Ulster for their valuable comments on the manuscript.

References

1. Walsh CP, Bestor TH. Cytosine methylation and mammalian development. Genes Dev 1999; 13:26-34. Available from: http://www.genesdev.org/cgi/content/full/13/1/26.
2. Herman JG, Umar A, Polyak K et al. Incidence and functional consequences of hMLH1 promoter hypermethylation in colorectal carcinoma. Proc Natl Acad Sci USA 1998; 95:6870-5. Available from: http://www.pnas.org/cgi/content/full/95/12/6870.
3. Yoder JA, Walsh CW, Bestor TH. Cytosine methylation and the ecology of intragenomic parasites. Trends in Genetics 1997; 13:335-340.
4. Xu GL, Bestor TH, Bourc'his D et al. Chromosome instability and immunodeficiency syndrome caused by mutations in a DNA methyltransferase gene. Nature 1999; 402:187-91.
5. Hansen RS, Wijmenga C, Luo P et al. The DNMT3B DNA methyltransferase gene is mutated in the ICF immunodeficiency syndrome. Proc Natl Acad Sci USA 1999; 96:14412-7.
6. Okano M, Bell DW, Haber DA et al. DNA methyltransferases Dnmt3a and Dnmt3b are essential for de novo methylation and mammalian development. Cell 1999; 99:247-57.
7. Gonzalo S, Jaco I, Fraga MF et al. DNA methyltransferases control telomere length and telomere recombination in mammalian cells. Nat Cell Biol 2006; 8:416-424.
8. Guo G, Wang W, Bradley A. Mismatch repair genes identified using genetic screens in blm-deficient embryonic stem cells. Nature 2004; 429:891-895.
9. Kim M, Trinh BN, Long TI et al. Dnmt1 deficiency leads to enhanced microsatellite instability in mouse embryonic stem cells. Nucleic Acids Res 2004; 32:5742-5749.
10. Wang KY, James Shen CK. DNA methyltransferase Dnmt1 and mismatch repair. Oncogene 2004; 23:7898-7902.
11. Kazazian HH Jr. Mobile elements: Drivers of genome evolution. Science 2004; 303:1626-1632.
12. Bourc'his D, Bestor TH. Meiotic catastrophe and retrotransposon reactivation in male germ cells lacking Dnmt3L. Nature 2004; 431:96-99.
13. Webster KE, O'Bryan MK, Fletcher S et al. Meiotic and epigenetic defects in Dnmt3L-knockout mouse spermatogenesis. Proc Natl Acad Sci USA 2005; 102:4068-4073.
14. Bender J. Cytosine methylation of repeated sequences in eukaryotes: The role of DNA pairing. Trends Biochem Sci 1998; 23:252-256.
15. Ginsburg M, Snow MH, McLaren A. Primordial germ cells in the mouse embryo during gastrulation. Development 1990; 110:521-528.
16. Hajkova P, Erhardt S, Lane N et al. Epigenetic reprogramming in mouse primordial germ cells. Mech Dev 2002; 117:15.
17. Li JY, Lees-Murdock DJ, Xu GL et al. Timing of establishment of paternal methylation imprints in the mouse. Genomics 2004; 84:952-960.
18. Yamazaki Y, Low EW, Marikawa Y et al. Adult mice cloned from migrating primordial germ cells. Proc Natl Acad Sci USA 2005; 102:11361-11366.
19. Tam PP, Zhou SX, Tan SS. X-chromosome activity of the mouse primordial germ cells revealed by the expression of an X-linked lacZ transgene. Development 1994; 120:2925-32.
20. Kerjean A, Couvert P, Heams T et al. In vitro follicular growth affects oocyte imprinting establishment in mice. Eur J Hum Genet 2003; 11:493-6.
21. Lees-Murdock D, De Felici M, Walsh C. Methylation dynamics of repetitive DNA elements in the mouse germ cell lineage. Genomics 2003; 82:230-237.
22. Mann MR, Lee SS, Doherty AS et al. Selective loss of imprinting in the placenta following preimplantation development in culture. Development 2004; 131:3727-3735.
23. Schumacher A, Doerfler W. Influence of in vitro manipulation on the stability of methylation patterns in the Snurf/Snrpn-imprinting region in mouse embryonic stem cells. Nucleic Acids Res 2004; 32:1566-1576.
24. Schmidt JV, Matteson PG, Jones BK et al. The Dlk1 and Gtl2 genes are linked and reciprocally imprinted. Genes Dev 2000; 14:1997-2002.
25. Kobayashi S, Wagatsuma H, Ono R et al. Mouse Peg9/Dlk1 and human PEG9/DLK1 are paternally expressed imprinted genes closely located to the maternally expressed imprinted genes: Mouse Meg3/Gtl2 and human MEG3. Genes Cells 2000; 5:1029-37.
26. Smit AFA. The origin of interspersed repeats in the human genome. Curr Op Genet Dev 1996; 6:743-748.
27. Hastie ND. Highly repeated DNA families in the genome of mus musculus. In: Lyon MF, Searle AG, eds. Genetic Variants and Strains of the Laboratory Mouse. Oxford: Oxford University Press, 1989:559-573.
28. Lucifero D, Mann MR, Bartolomei MS et al. Gene-specific timing and epigenetic memory in oocyte imprinting. Hum Mol Genet 2004; 13:839-849.
29. Howlett SK, Reik W. Methylation levels of maternal and paternal genomes during preimplantation development. Development 1991; 113:119-27.

30. Lane N, Dean W, Erhardt S et al. Resistance of IAPs to methylation reprogramming may provide a mechanism for epigenetic inheritance in the mouse. Genesis 2003; 35:88-93.
31. Sado T, Fenner MH, Tan SS et al. X inactivation in the mouse embryo deficient for Dnmt1: Distinct effect of hypomethylation on imprinted and random X inactivation. Dev Biol 2000; 225:294-303.
32. Monk M, McLaren A. X-chromosome activity in foetal germ cells of the mouse. J Embryol Exp Morphol 1981; 63:75-84.
33. McLaren A, Monk M. X-chromosome activity in the germ cells of sex-reversed mouse embryos. J Reprod Fertil 1981; 63:533-7.
34. Boumil RM, Ogawa Y, Sun BK et al. Differential methylation of xite and CTCF sites in tsix mirrors the pattern of X-inactivation choice in mice. Mol Cell Biol 2006; 26:2109-2117.
35. Wolffe AP, Jones PL, Wade PA. DNA demethylation. Proc Natl Acad Sci USA 1999; 96:5894-5896.
36. Bird A. DNA methylation patterns and epigenetic memory. Genes Dev 2002; 16:6-21.
37. Walsh CP, Xu GL. Cytosine methylation and DNA repair. Curr Top Microbiol Immunol 2006; 301:283-315.
38. Gehring M, Huh JH, Hsieh TF et al. DEMETER DNA glycosylase establishes MEDEA polycomb gene self-imprinting by allele-specific demethylation. Cell 2006; 124:495-506.
39. Morgan HD, Dean W, Coker HA et al. Activation-induced cytidine deaminase deaminates 5-methyl-cytosine in DNA and is expressed in pluripotent tissues: Implications for epigenetic reprogramming. J Biol Chem 2004; 279:52353-52360.
40. Ginsburg M, Snow MH, McLaren A. Primordial germ cells in the mouse embryo during gastrulation. Development 1990; 110:521-8.
41. Sato Y, Terada Y, Utsunomiya H et al. Immunohistochemical localization of steroidogenic enzymes in human follicle following xenotransplantation of the human ovarian cortex into NOD-SCID mice. Mol Reprod Dev 2003; 65:67-72.
42. Reik W, Walter J. Evolution of imprinting mechanisms: The battle of the sexes begins in the zygote. Nat Genet 2001; 27:255-6.
43. Davis TL, Trasler JM, Moss SB et al. Acquisition of the H19 methylation imprint occurs differentially on the parental alleles during spermatogenesis. Genomics 1999; 58:18-28.
44. Davis TL, Yang GJ, McCarrey JR et al. The H19 methylation imprint is erased and re-established differentially on the parental alleles during male germ cell development. Hum Mol Genet 2000; 9:2885-94.
45. Ueda T, Abe K, Miura A et al. The paternal methylation imprint of the mouse H19 locus is acquired in the gonocyte stage during foetal testis development. Genes Cells 2000; 5:649-659.
46. Shamanski FL, Kimura Y, Lavoir MC et al. Status of genomic imprinting in mouse spermatids. Hum Reprod 1999; 14:1050-1056.
47. Bao S, Obata Y, Carroll J et al. Epigenetic modifications necessary for normal development are established during oocyte growth in mice. Biol Reprod 2000; 62:616-621.
48. Obata Y, Kono T. Maternal primary imprinting is established at a specific time for each gene throughout oocyte growth. J Biol Chem 2002; 277:5285-9.
49. Lucifero D, Mertineit C, Clarke HJ et al. Methylation dynamics of imprinted genes in mouse germ cells. Genomics 2002; 79:530-8.
50. Hiura H, Obata Y, Komiyama J et al. Oocyte growth-dependent progression of maternal imprinting in mice. Genes Cells 2006; 11:353-361.
51. Walsh CP, Chaillet JR, Bestor TH. Transcription of IAP endogenous retroviruses is constrained by cytosine methylation. Nat Genet 1998; 20:116-7.
52. Lees-Murdock DJ, Walsh CP. Developmental regulation of DNA methyltransferases. Available at: http://www.interscience.wiley.com/mrw/eggpb
53. Lees-Murdock DJ, Shovlin TC, Gardiner T et al. DNA methyltransferase expression in the mouse germ line during periods of de novo methylation. Dev Dyn 2005; 232:992-1002.
54. Thorvaldsen JL, Verona RI, Bartolomei MS. X-tra! X-tra! News from the mouse X chromosome. Dev Biol. In Press.
55. Nguyen DK, Disteche CM. Dosage compensation of the active X chromosome in mammals. Nat Genet 2006; 38:47-53.
56. Mayer W, Niveleau A, Walter J et al. Demethylation of the zygotic paternal genome. Nature 2000; 403:501-502.
57. Oswald J, Engemann S, Lane N et al. Active demethylation of the paternal genome in the mouse zygote. Curr Biol 2000; 10:475-478.
58. Olek A, Walter J. The pre-implantation ontogeny of the H19 methylation imprint [letter]. Nat Genet 1997; 17:275-6.
59. Howell CY, Bestor TH, Ding F et al. Genomic imprinting disrupted by a maternal effect mutation in the Dnmt1 gene. Cell 2001; 104:829-38.

60. Kaslow DC, Migeon BR. DNA methylation stabilizes X chromosome inactivation in eutherians but not in marsupials: Evidence for multistep maintenance of mammalian X dosage compensation. Proc Natl Acad Sci USA 1987; 84:6210-6214.
61. Hansen RS. X inactivation-specific methylation of LINE-1 elements by DNMT3B: Implications for the lyon repeat hypothesis. Hum Mol Genet 2003; 12:2559-2567.
62. Chen RZ, Pettersson U, Beard C et al. DNA hypomethylation leads to elevated mutation rates. Nature 1998; 395:89-93.
63. Yoder JA, Soman N, Verdine GV et al. DNA methyltransferases in mouse tissues and cells: Studies with a mechanism-based probe. J Mol Biol 1997 (in press).
64. Fedoriw AM, Stein P, Svoboda P et al. Transgenic RNAi reveals essential function for CTCF in H19 gene imprinting. Science 2004; 303:238-240.
65. Yoon BJ, Herman H, Sikora A et al. Regulation of DNA methylation of Rasgrf1. Nat Genet 2002; 30:92-6.
66. Herman H, Lu M, Anggraini M et al. Trans allele methylation and paramutation-like effects in mice. Nat Genet 2003; 34:199-202.
67. Reinhart B, Eljanne M, Chaillet JR. Shared role for differentially methylated domains of imprinted genes. Mol Cell Biol 2002; 22:2089-2098.
68. Lyon MF. The lyon and the LINE hypothesis. Semin Cell Dev Biol 2003; 14:313-318.
69. Lyon MF. X-chromosome inactivation: A repeat hypothesis. Cytogenet Cell Genet 1998; 80:133-137.
70. Bailey JA, Carrel L, Chakravarti A et al. Molecular evidence for a relationship between LINE-1 elements and X chromosome inactivation: The lyon repeat hypothesis. Proc Natl Acad Sci USA 2000; 97:6634-6639.
71. Parish DA, Vise P, Wichman HA et al. Distribution of LINEs and other repetitive elements in the karyotype of the bat carollia: Implications for X-chromosome inactivation. Cytogenet Genome Res 2002; 96:191-197.
72. Li E, Beard C, Jaenisch R. Role for DNA methylation in genomic imprinting. Nature 1993; 366:362-365.
73. Bell AC, Felsenfeld G. Methylation of a CTCF-dependent boundary controls imprinted expression of the Igf2 gene. Nature 2000; 405:482-5.
74. Hark AT, Schoenherr CJ, Katz DJ et al. CTCF mediates methylation-sensitive enhancer-blocking activity at the H19/Igf2 locus. Nature 2000; 405:486-9.
75. Csankovszki G, Nagy A, Jaenisch R. Synergism of xist RNA, DNA methylation and histone hypoacetylation in maintaining X chromosome inactivation. J Cell Biol 2001; 153:773-84.

CHAPTER 2

Control of Imprinting at the *Gnas* Cluster

Jo Peters* and Christine M. Williamson

Abstract

Genomic imprinting is a form of epigenetic regulation in mammals whereby a small subset of genes is silenced according to parental origin. Early work had indicated regions of the genome that were likely to contain imprinted genes. Distal mouse chromosome 2 is one such region and is associated with devastating but ostensibly opposite phenotypes when exclusively maternally or paternally derived. Misexpression of proteins encoded at the *Gnas* complex, which is located within the region, can largely account for the imprinting phenotypes. *Gnas* is a complex locus with extraordinary transcriptional and regulatory complexity. It gives rise to alternatively spliced isoforms that show maternal-, paternal- and biallelic expression as well as a noncoding antisense transcript. The objective of our work at Harwell is to unravel mechanisms controlling the expression of these transcripts. We have performed targeted deletion analysis to test candidate regulatory regions within the *Gnas* complex and, unlike other imprinted domains, two major control regions have been identified. One controls the imprinted expression of a single transcript and is subsidiary to and must interact with, a principal control region that affects the expression of all transcripts. This principal region contains the promoter for the antisense transcript, expression of which may have a major role in controlling imprinting at the *Gnas* cluster.

Background

The imprinting region in distal chromosome 2 of the mouse was one of the first to be described.[1] Mice with two maternally derived copies of distal chromosome 2, but no paternally derived copies (MatDp(dist2)), had long thin bodies, failed to suckle, became inert and died within a few hours of birth. On the other hand, mice with two paternally derived copies of distal chromosome 2 but no maternally derived copies (PatDp(dist2)), had an apparently opposite phenotype, for they had short square bodies, were oedematous, notably hyperactive and died within a few days of birth. Using genetic approaches the limits of the region were defined by the chromosome 2 breakpoints in 2H3 and 2H4 in the reciprocal translocations T(2;8)2Wa and T(2;16)28H.[2,3] The region was estimated to be 7 Mb in size (Evans EP (pers.comm.)) and contained the *Gnas* locus.[4] From clinical and biochemical studies in humans there was evidence that the orthologous *GNAS* locus was imprinted and it had been inferred that the paternally derived allele was silenced in renal proximal tubules.[5] Williamson et al[4] provided the first evidence of imprinting at the mouse *Gnas* locus, but, surprisingly, found that the maternally derived allele, not the paternally derived allele was repressed. Also, repression occurred in the glomerulus, not the proximal tubules. This finding was probably confounded by the complexity of the locus (Fig. 1), which was completely

---

*Corresponding Author: Jo Peters—MRC Mammalian Genetics Unit, Harwell, Oxfordshire, OX11 0RD, U.K. Email: j.peters@har.mrc.ac.uk

Genomic Imprinting, edited by Jon F. Wilkins. ©2008 Landes Bioscience and Springer Science+Business Media.

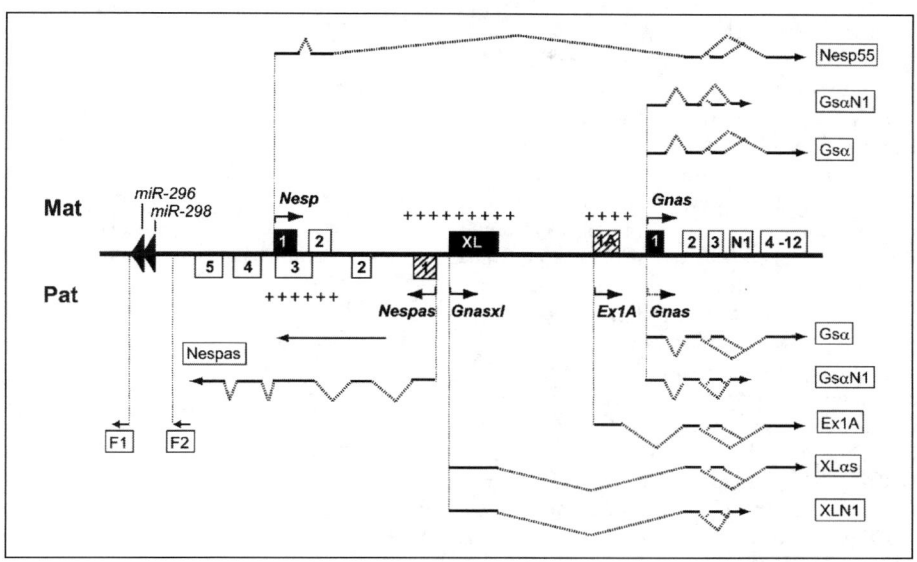

Figure 1. Genomic organisation of the *Gnas* cluster in the mouse. Features of the maternal (Mat) and paternal (Pat) allele are shown above and below the line, respectively. The arrows show initiation and direction of transcription. Transcription of *Gnas* from the paternal allele is shown as a dotted line to indicate that its promoter is inactive in some tissues. Although at least 50 different transcripts have been identified in the cluster,[14] for simplicity only the splice variants relevant to the text are shown above and below the line. The first exons of the protein coding transcripts are shown as filled rectangles and the first exons of the noncoding transcripts are shown as striped boxes. The *Nespas* and *Exon 1A* transcripts are noncoding. Maternally and paternally methylated regions are shown by + symbols above and below the line, respectively. The filled triangles on the line, *miR-296* and *miR-298* are experimentally detected small RNA genes;[19] their imprinting status has not been determined. F1 and F2 are FANTOM2 cDNA clones that are from paternally expressed transcripts on the antisense strand.[14] The figure is not to scale (adapted from Plagge et al[26] and Peters et al[7]).

unknown at the time. Subsequently, analysis of mice with a null allele of *Gnas* showed imprinted expression at the locus.[6] The paternal allele was silent in renal cortex and also in brown and white adipose tissue.[6] The phenotype of mice with a maternally derived null allele appeared to show similarities to that of PatDp(dist2), which lack a maternal copy, whereas the phenotype of mice with a paternally derived null allele showed similarities to that seen in MatDp(dist2), which lack a paternal copy. These observations imply that misexpression of the *Gnas* locus could account for much of the phenotype in both PatDp(dist2) and MatDp(dist2).

Extent of the Region

Expression of genes in a 1 Mb region around *Gnas* and the human ortholog, *GNAS* has been examined, but only *Gnas/GNAS* showed imprinted expression.[7] Gene order is conserved in humans and mice across the region tested. So far there is no evidence for any other imprinted loci within the distal chromosome 2 imprinting region. The imprinted domain in mouse distal chromosome 2 may therefore be restricted to the *Gnas* locus and only be about 60 kb in size, which is compact in comparison to some other imprinted domains. The size of the human *GNAS* locus is of a similar order (around 70 kb), but in humans there also appears to be a *cis*-acting imprinting control element at *STX16* around 200 kb upstream of *GNAS*.[8,9]

Transcripts at the *Gnas* Locus

The compact imprinted *Gnas* locus is highly complex (Fig. 1). It gives rise to maternally, paternally and biallelically expressed transcripts that share a common set of downstream exons.[10] These transcripts arise from four alternative first exons and promoters that splice on to exon 2 of *Gnas*. To add to the confusion, the term *Gnas* is used for one of the transcripts within it as well as for the whole locus. The furthest upstream first exon is *Nesp* exon 1, which lies 45.7 kb upstream of *Gnas* exon 1 and gives rise to a maternally expressed *Nesp* transcript.[10,11] Next is the XL exon, which lies approximately 32 kb upstream of *Gnas* exon 1 and gives rise to a paternally expressed *Gnasxl* transcript.[10,11] The 1A exon lies 2.3 kb from *Gnas* exon 1 and gives rise to a noncoding paternally expressed *Exon 1A* transcript.[12] Lastly, the *Gnas* transcript, arising from *Gnas* exon 1, is biallelically expressed in most tissues but is predominantly maternally expressed in a few.[6,13] Transcripts arising from the *Nesp*, *Gnasxl*, *Exon 1A* and *Gnas* promoters that terminate upstream of *Gnas* exon 2 have also been identified.[14] In addition, *Nespas* exon 1 gives rise to a paternally expressed noncoding transcript that is transcribed antisense to *Nesp*[15,16,17] and starts 2.4 kb upstream of the *Gnasxl* translation initiation site.[18] *Nespas* transcripts cover more than 30 kb of genomic DNA[14] and exist as spliced and unspliced forms.[17] Also, two microRNAs *miR-296* and *miR-298* map to the *Gnas* locus (RJ Holmes, pers. comm.).[19] They lie at the 3' end of *Nespas* between paternally expressed F1 and F2 cDNAs (RJ Holmes, pers. comm.).[14]

The promoters for *Nesp*, *Gnasxl*, *Exon 1A* and *Nespas* all lie within one of three differentially methylated regions (DMRs). The exception is *Gnas*, which is unmethylated on both parental alleles. Of the three DMRs, one, established postfertilization is paternally methylated and two, established during gametogenesis, are maternally methylated. The paternally methylated region covers the promoter for the maternally expressed *Nesp* transcript.[10,11] One extensive maternally methylated region covers the promoters for the paternally expressed *Nespas* and *Gnasxl* transcripts,[20] and a second maternally methylated region covers the promoter for the paternally expressed *Exon 1A* transcript.[12]

Proteins Encoded at the *Gnas* Complex Locus

Several proteins are encoded by imprinted transcripts at the *Gnas* locus (Fig. 1). The transcript from the *Gnas* promoter extends from *Gnas* exons 1 to 12 and encodes Gsα, the alpha stimulatory subunit of the widely expressed heterotrimeric protein, Gs that is required for hormone stimulated cAMP production. A shortened transcript, GsαN1, that has a terminal exon, N1, in intron 3 is abundant in brain.[21] Gsα is mainly biallelically expressed, but shows preferential expression from the maternal allele in some tissues in the mouse such as renal proximal tubules and adipose tissue.[6]

The *Gnasxl* promoter drives transcription of paternally expressed transcripts that encode several different proteins. One transcript extends from the *Gnasxl* XL exon through *Gnas* exons 2 to 12 encoding XLαs, an isoform of Gsα in which the N-terminal of Gsα is replaced by a large acidic domain encoded by the XL exon.[22] XLαs has much in common with Gsα, including heterotrimer formation and activation of adenylyl cyclase leading to cAMP production.[22,23] Activation of adenylyl cyclase can occur via coupling of XLαs to receptors that typically couple with Gs in transfected cells.[24] A shortened transcript with a terminal exon N1 encodes a protein XLN1.[25] The XLαs and XLN1 transcripts are strongly expressed in neuroendocrine tissues in the adult,[25] and the XLαs transcript is highly expressed in brown and white adipose tissue in the perinatal period.[26] There is an alternative open reading frame (ORF) within the XL exon encoding ALEX, expressed in neuroendocrine cells that can interact with XLαs.[27]

A maternally expressed transcript from the *Nesp* promoter encodes a chromogranin-like protein, Nesp55, of unknown function, that is associated with the constitutive secretory pathway.[28] Nesp55 is expressed in neural tissue and the adrenal medulla.[29,30] The Nesp55 ORF lies entirely in the second *Nesp* exon. For further discussion of the possible function of Nesp55, see the chapter by Davies et al.

Function of Proteins Encoded at the *Gnas* Locus

The role of proteins encoded by the *Gnas* locus has been revealed from studies of mutants. Loss of function of either *Gnasxl* or *Gnas* results in severe, but different phenotypes in the perinatal period.

Mice with paternally inherited null mutations resulting in loss of XLαs, XLN1 and presumably ALEX as well, are poor at suckling, show decreased adiposity and activity and generally die within the perinatal period.[6,26,31] This phenotype is remarkably similar to that seen in MatDp(dist2) mice. The suckling defect may well be due to loss of XLN1. Paternal inheritance of the missense *Gnas* mutation *Oed-Sml* affects XLαs but not XLN1 and does not affect suckling (JA Skinner and J Peters, unpublished), but causes postnatal growth retardation.[32]

Mice with maternal inheritance of mutations resulting in loss of Gsα and GsαN1 show sub-cutaneous oedema that is marked in the perinatal period, have increased adiposity and generally die prior to weaning.[6,33,34] Maternal inheritance of the *Oed-Sml* mutation that affects Gsα, but not GsαN1, results in newborns with very marked sub-cutaneous oedema and increased adiposity, with most dying before weaning (Skinner and Peters, unpublished).[32] This phenotype shows marked similarities to that seen in PatDp(dist2) mice. Maternal inheritance of a loss-of-function mutation in Nesp55 does not result in a phenotype in the preweaning period.[29] Thus one conclusion is that mis-expression of *Gnas* can account for much of the phenotype in PatDp(dist2) and mis-expression of *Gnasxl* can account for much of the phenotype in MatDp(dist2).

Biochemical and metabolic studies show that normal functions of XLαs encoded by *Gnasxl* must be to promote growth, increase fat mass and lipid accumulation, lessen metabolic rate, elevate serum glucose, insulin and triglycerides; whereas normal functions of Gsα encoded by *Gnas* must be to diminish growth, decrease fat mass and lipid accumulation, elevate metabolic rate and lessen serum glucose, insulin and triglycerides.[26,33-37] There is also evidence that XLαs can depress cAMP formation and can act antagonistically to Gsα.[26] A second conclusion is that XLαs and Gsα, exert major, but opposite effects on development after birth. The metabolic consequences of gene expression at the *Gnas* locus are discussed further in the chapter by Frontera et al.

Gnas is the first example of antagonism between maternally and paternally expressed proteins encoded at the same locus. This has relevance for the kinship theory for the evolution of imprinting (see the chapter by Moore and Mills). According to this theory, imprinting has evolved because of an evolutionary conflict in individuals between maternally and paternally derived alleles. It is in the interests of paternally expressed genes to acquire resources from the mother and promote growth, whereas it is in the interests of maternally expressed genes to restrict demand on maternal resources and restrict growth.[38] Paternally expressed XLαs is directly involved in the acquisition of maternal resources, in the form of milk. It is therefore to the advantage of the paternal allele to ensure that *Gnasxl* is expressed but *Gnas* is repressed, but to the advantage of the maternal allele to ensure that *Gnas* is expressed but *Gnasxl* is repressed. There is an intricate set of regulatory controls to ensure paternal expression of *Gnasxl* and maternal expression of *Gnas*.

Imprinting Centers

Imprinted genes tend to occur in clusters. The imprinting of genes within a cluster is regulated by a region called an imprinting center, or imprinting control region (ICR). An imprinting center can act over hundreds of kilobases to co-ordinately regulate the expression of many genes. Imprinting centers have been identified at eight imprinting clusters in six chromosome regions (see review by Lewis and Reik),[39] and most have been established by in vivo analysis of targeted deletions. There are a number of characteristics common to imprinting centers. Imprinting centers are CpG-rich regions or CpG islands; are necessary for imprinted expression; regulate the imprinted expression of other genes in the cluster *in cis*; are differentially methylated in the germlines with the differential methylation being maintained in somatic tissues in the offspring; may show additional differential chromatin modifications according to parental origin; can act as chromatin insulators; contain the promoters for noncoding RNAs.[39,40]

The *Gnas* cluster is small but contains two regions with the characteristics of an ICR (Fig. 2). These are CpG island regions that are differentially methylated in the germlines. This differential methylation is maintained in the somatic tissues of the offspring.

One of these regions is at *Exon* 1A, the *Exon* 1A DMR and is unmethylated on the paternally derived allele but methylated on the maternally derived allele.[12] It is approximately 2.5 kb in size,

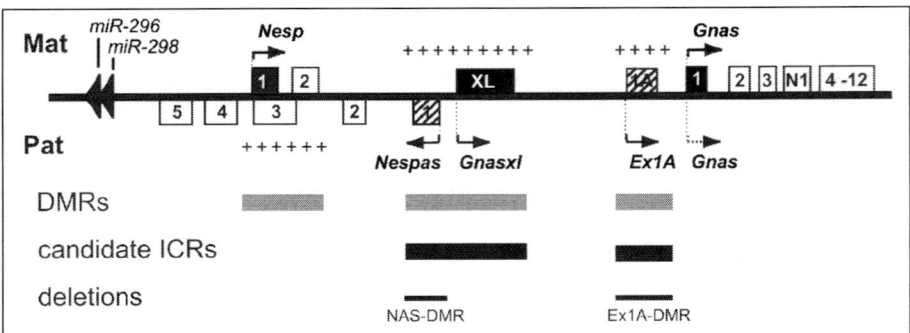

Figure 2. Two candidate imprinting control regions (ICRs) in the *Gnas* cluster. The features are depicted as in Figure 1. The three DMRs are shown as grey filled rectangles and the two candidate ICRs are shown as black filled rectangles. The two candidate ICRs are germline maternally methylated regions that are associated with the start of noncoding RNAs whereas the other DMR at *Nesp* is established after fertilisation. The black lines show the position of the deleted regions, designated NAS-DMR and Ex1A-DMR. The figure is not to scale.

covers the *Exon* 1A exon and contains the promoter for a noncoding RNA, the *Exon* 1A transcript.[12] The 3' end of the *Exon* 1A DMR is about 1 kb upstream of the 5' end of exon 1 of *Gnas*.[12] It shows parental specific histone modifications that correlate with the DNA methylation status; the maternally derived *Exon 1A* DMR is marked by repressive H3-K9 methylation characteristic of heterochromatin, whereas the paternally derived *Exon 1A* DMR is associated with H3-H4 acetylation and H3-K4 methylation, characteristic of euchromatin.[41]

The *Nespas* and *Gnasxl* promoters are embedded in the second germline differentially methylated CpG island. This is extensive, covering more than 5.8 kb.[20] It covers the promoter for a noncoding antisense RNA, *Nespas*, as well as the *Gnasxl* promoter region and contains DNase I hypersensitive sites, indicative of an open chromatin structure, that are specific to the hypomethylated paternally derived allele.[20] MacroH2A1, the variant form of histone H2A is found preferentially on the hypermethylated maternal allele in the *Nespas* promoter region.[42] Often tandem arrays of transcription factor binding sites are found at ICRs and seven copies of the potential binding sites for the transcription factor YY1 have been found in the DMR in intron 1 of *Nespas*.[43]

There is a third DMR at the *Gnas* locus. A paternally methylated region covers the promoter for *Nesp* and extends over 4.4 kb and includes both *Nesp* exons.[10,11,20] As this paternal methylation is established post fertilization,[12] this DMR is not a candidate for an ICR. Studies of human patients with pseudohypoparathyroidism type Ib (PHP1b) have shown that there are *cis*-acting elements at both the NESP55 DMR and 200kb upstream at the *STX16* locus that are essential for methylating the maternally derived allele.[8,9,44,45] For more on pseudohypoparathyroidism, see the chapter by Bastepe.

The occurrence of two regions in the mouse *Gnas* cluster with the characteristics of imprinting centers raised the possibility that the cluster could be divided into two domains whose imprinting was regulated independently by separate elements. In humans there was some evidence indicating that the exon A/B DMR (the human homologue of the mouse *Exon 1A* DMR) controlled the imprinted expression of *GNAS*. Most patients with the autosomal dominant form of pseudohypoparathyroidism type Ib have lost methylation of the exon A/B DMR alone.[8,46] It has been suggested that the unmethylated exon A/B DMR suppresses *GNAS* expression. Thus in patients with PHP Ib loss of maternal methylation of exon A/B is postulated to lead to biallelic loss of *GNAS* expression in renal proximal tubules and PTH resistance.

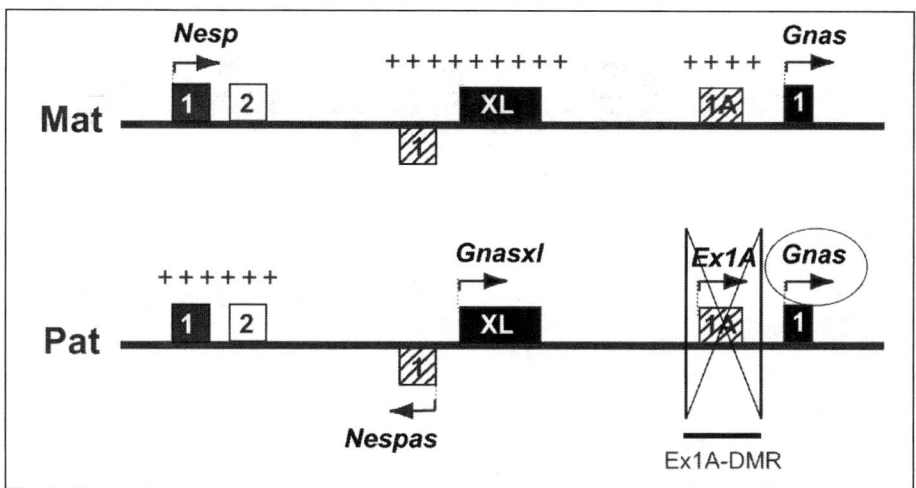

Figure 3. A unique ICR at the *Exon* 1A DMR specifically regulates the tissue-specific imprinted expression of the *Gnas* gene. The features are depicted as in Figure 1 except that the maternally and paternally inherited chromosomes are separate; sense and antisense transcription are shown above and below the lines, respectively. Effects due to the presence of the deletion are circled (adapted from Williamson et al[13]).

The *Exon 1A* DMR Controls the Expression of *Gnas*

In order to test directly the role of the *Exon 1A* DMR in the mouse, Williamson et al[13] made a 2.3 kb targeted deletion, designated the Ex1A-DMR in the mouse covering most of the DMR (Fig. 3). On paternal inheritance of the deletion *Gnas* was derepressed on the paternal allele in brown fat, a tissue in which *Gnas* is normally preferentially maternally expressed. Furthermore, Gsα-mediated PTH signaling, an indicator of *Gnas* expression in renal proximal tubules, was increased. This finding implies that loss of silencing of the paternal *Gnas* allele in renal proximal tubules had occurred. The imprinted expression of other transcripts in the cluster, *Nesp*, *Gnasxl* and *Nespas* in the cluster was unaffected. These data show that the *Exon 1A* DMR is sufficient for imprinted expression of *Gnas* and must contain element(s) controlling the imprinted expression of *Gnas* alone. These results were confirmed following analysis of a 4.7 kb targeted deletion of the *Exon 1A* DMR by Chen et al[47] who showed that on paternal inheritance *Gnas* expression in renal proximal tubules was significantly elevated.[47] The raised expression was attributed to derepression of *Gnas* on the paternal allele. Furthermore the methylation of both the germline *Nespas* DMR and the *Nesp* DMR was unaffected.[47] Taking the results of expression and methylation studies together, the *Exon 1A* DMR controls the imprinted expression of *Gnas* alone. Thus there had to be a second ICR regulating the imprinted expression of the other transcripts in the cluster.

The *Nespas* DMR is the Principal ICR in the *Gnas* Cluster

The germline *Nespas* DMR was a good candidate because, like ICRs in the *Kcnq1*, *Igf2r* and PWS clusters, it contains the promoter for an antisense noncoding transcript. A 1.6 kb targeted deletion, designated the NAS-DMR covering the *Nespas* promoter, first exon and some intronic sequence was made (Fig. 4).[48] On paternal inheritance the expression of all the major transcripts in the cluster was affected. As expected, because the *Nespas* promoter had been deleted, *Nespas* was no longer expressed. The *Nesp* transcript that is usually repressed on the paternal allele became derepressed. Thus on paternal inheritance of the deletion *Nesp* became biallelic. Expression of *Gnasxl* was diminished, probably because a regulatory element required for its expression lay within the deleted region. Expression of *Exon 1A* was also diminished, whereas there was increased

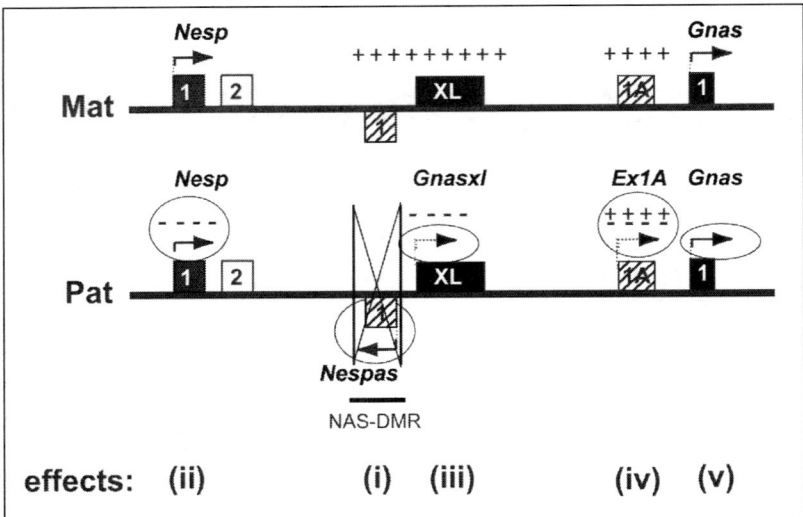

Figure 4. NAS-DMR is the principal ICR at the *Gnas* cluster. Paternal inheritance of the NAS-DMR deletion caused: (i) loss of *Nespas* transcription, (ii) derepression of *Nesp* on the paternal allele associated with loss of methylation at *Nesp* on the paternal allele, (iii) a reduction in *Gnasxl* level from the paternal allele despite the allele remaining unmethylated, (iv) a reduction in *Exon 1A* level from the paternal allele associated with a gain of methylation and (v) an upregulation of *Gnas* from the paternal allele in brown fat. Effects due to the presence of the deletion are circled (adapted from Williamson et al[48]).

expression of *Gnas* from the paternal allele in brown fat, a tissue in which *Gnas* is usually repressed on the paternal allele. Methylation of both the *Nesp* DMR and the *Exon 1A* DMR was affected. The *Nesp* DMR, normally methylated on the paternal allele, had lost methylation on paternal inheritance of the deletion whereas the *Exon 1A* DMR, normally unmethylated on the paternal allele, had gained some methylation on paternal inheritance of the deletion.

Thus the *Nespas* DMR has the characteristics of an ICR and affects the imprinted expression of all major transcripts in the *Gnas* cluster. Therefore it can be concluded that it is the principal ICR for the cluster and the *Exon 1A* DMR is a subsidiary control element affecting the expression of *Gnas*. It can also be concluded that there is a single imprinted domain at the *Gnas* cluster.

Interaction between the *Nespas* DMR and the *Exon 1A* DMR

The *Nespas* DMR must influence the methylation of the *Exon 1A* DMR and a normal function of the *Nespas* DMR on the paternal allele must be to protect the *Exon 1A* DMR from methylation. This hierarchical arrangement of DMRs is reminiscent of the *Igf2/H19* cluster, where the imprinting control region, the *H19* DMR, when unmethylated on the maternal allele, protects the upstream *Igf2* DMRs 1 and 2 from methylation.[49] It was proposed that the protection occurred via interaction between the DMRs. Subsequently it was shown that the *Igf2/H19* DMRs interact resulting in the formation of parental-specific chromatin loops.[50] It is unknown how the *Nespas* DMR interacts with the *Exon* 1A DMR, but chromatin looping is one possibility.

From mouse studies and studies of human patients with PHP1b there is evidence that loss of methylation of the *Exon 1A* DMR is associated with repression of *Gnas* from the paternal allele in some tissues (see the chapter by Bastepe). It is entirely consistent to find that on paternal inheritance of the *Nespas* deletion, increased methylation of the *Exon 1A* DMR is associated with increased expression of *Gnas*. Thus it appears that on the paternal allele the *Nespas* DMR protects

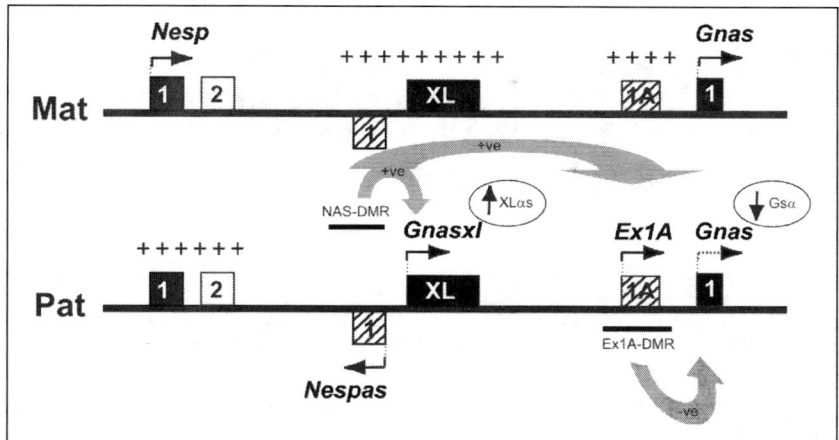

Figure 5. Strategies of the paternal chromosome to regulate Gsα levels. Although the mechanism of action of the NAS-DMR and the Ex1A-DMR is unknown, the grey arrows labelled +ve show the NAS-DMR is required on the paternal allele to enhance XLαs and Ex1A. The NAS-DMR also interacts with the Ex1A-DMR to downregulate Gsα levels (shown as grey arrow labelled –ve).

the the *Exon 1A* DMR from methylation and in turn the unmethylated *Exon 1A* represses *Gnas* expression (Fig. 5). We can conclude that the imprinted expression of *Gnas* on the paternal allele is controlled by two DMRs.

The Role of *Nespas*

Like other ICRs that acquire a methylation imprint during oogenesis, the *Nespas* ICR contains the promoter for a noncoding RNA that is transcribed antisense with respect to a protein-coding gene within the cluster. The proximal chromosome 17 *Igf2r/Air* ICR contains the promoter for *Air*, a well characterized RNA that is transcribed antisense to *Igf2r*;[51] the distal chromosome 7 IC2 ICR contains the promoter for *Kcnq1ot1*, an RNA that is transcribed antisense to *Kcnq1*,[52] and the promoter for *Ube3a-ats*, an RNA that is transcribed antisense to *Ube3a*, is near the central chromosome 7 PWS ICR.[53] Expression of these noncoding RNAs on the paternally derived allele is associated with repression of protein coding genes in *cis*. This suggests that the noncoding RNAs have a silencing function, a suggestion supported by the finding that deletion of the promoters for these noncoding RNAs is associated with derepression of protein coding genes in *cis*.[54,55] Both *Air* and *Kcnq1ot1* RNAs have been shown to have silencing function over the whole of their respective imprinting domains and are bi-directional *cis*-acting domain silencers.[56,57] The mechanisms whereby *Air* and *Kcnqot1* act as silencers are unknown, but it is proposed that either the noncoding RNA transcript itself or the act of transcription is important in the silencing process. The similarities in the *Nespas* ICR and both the *Igf2r/Air* ICR and the IC2 ICR have raised the prospect that *Nespas* expression could also have a silencing function. However, although expression of *Nespas* is associated with repression of the protein-coding *Nesp* transcript, it is also associated in *cis* with expression, not repression, of the protein-coding transcript *Gnasxl*. Thus, unlike *Air* or *Kcnq1ot1*, *Nespas* appears to have silencing function over only part of the *Gnas* imprinting domain.

The Role of the *Exon 1A* DMR

The mechanism whereby *Exon 1A* represses expression of *Gnas* on the paternal allele in specific tissues is unknown, but two models have been proposed.[7,58] In one, the repressor model, it has been proposed that a silencer protein binds to the unmethylated *Exon 1A* DMR on the paternal allele.

Once bound, a repressive chromatin domain is set up, preventing transcription of *Gnas* exon 1. The expression of the silencer protein must be restricted to tissues in which *Gnas* shows imprinted expression and the silencer protein must only bind to unmethylated DNA. Thus on the maternal allele where *Exon 1A* is methylated, the silencer protein is unable to bind, the chromatin remains in an open state and *Gnas* is expressed. On paternal inheritance of the deletion of the *Exon 1A* DMR the binding sites for the silencer protein would be lost, the chromatin would remain open and *Gnas* would be derepressed on the paternal allele. On maternal inheritance of the deletion the expression of *Gnas* would be unaffected. In the second model, the insulator model, it has been proposed that the unmethylated *Exon 1A* DMR is an insulator that binds a protein such as CTCF on the paternal allele. Access of tissue-specific enhancers upstream of the *Exon 1A* DMR to the *Gnas* promoter would be blocked and *Gnas* would be silenced on the paternal allele. On the maternal allele the *Exon 1A* DMR is methylated and unable to bind an insulator protein. Therefore upstream enhancers can access the *Gnas* promoter and *Gnas* is expressed from the maternal allele. The positions of enhancers for transcription within the *Gnas* complex are unknown at this time. On paternal inheritance of the deletion of the *Exon 1A* DMR, the insulator protein would have nothing to bind to, the upstream enhancers would have access to the *Gnas* promoter and *Gnas* would be derepressed on the paternal allele. On maternal inheritance of the deletion the expression of *Gnas* would be unaffected. The expression of *Gnas* on maternal and paternal inheritance of the *Exon 1A* DMR deletion accords with the predictions of each of these models.

On paternal inheritance of the *Nespas* DMR deletion there is some increase in the methylation of the paternally derived *Exon 1A* DMR and a considerable increase in *Gnas* expression from the paternal allele. It is to be expected that binding of a silencer or insulator protein would be severely compromised when the DMR is partially methylated Alternatively, there may be two populations of cells, in one of which the *Exon 1A* DMR is completely unmethylated and can bind a silencer or insulator protein, but in the other population the *Exon 1A* DMR is completely methylated and incapable of protein binding.

It had seemed unlikely that transcription of *Exon 1A* or the *Exon 1A* transcript could have a role in silencing *Gnas* on the paternal allele because both *Exon 1A* and *Gnas* transcripts are expressed ubiquitously but silencing of *Gnas* only occurs in a few tissues. However, the findings that (1) there is an inverse relationship in the level of expression of *Exon 1A* and *Gnas* in brown fat on paternal inheritance of the *Nespas* DMR deletion and (2) *Exon 1A* is highly expressed in brown fat, a tissue in which *Gnas* shows imprinted expression, but poorly expressed in liver, a tissue in which *Gnas* does not show imprinted expression, suggest that expression of *Exon 1A* could be important in modulating *Gnas* expression. *Exon 1A* expression could repress *Gnas* expression by an RNA mediated model, by promoter competition, or by transcriptional interference.

Conclusions

Gnas is a complex imprinted locus that encodes at least three proteins, two of which, XLαs and Gsα, act antagonistically and have important roles in development after birth. We conclude that imprinting has occurred at this locus to ensure paternal specific expression of XLαs and, in some tissues, maternal-specific expression of Gsα. An intricate system of regulatory controls has evolved to enable imprinted expression of XLαs and Gsα. The maternally methylated region at *Exon 1A*, the *Exon 1A* DMR, controls only the imprinted expression of Gsα. A second maternally methylated region covering the promoter of the antisense transcript *Nespas*, the *Nespas* DMR, affects the imprinted expression of all major transcripts at the *Gnas* locus and is the principal controlling region. The *Nespas* DMR must interact with the subsidiary *Exon 1A* DMR to control the imprinted expression of Gsα. Future challenges are to establish how the *Nespas* DMR and the *Exon 1A DMR* interact and how the *Exon 1A* DMR controls Gsα

Acknowledgement

We thank Bruce Cattanach for valuable comments on the manuscript.

References

1. Cattanach BM, Kirk M. Differential activity of maternally and paternally derived chromosome regions in mice. Nature 1985; 315:496-498.
2. Cattanach BM, Evans EP, Burtenshaw MD et al. Further delimitation of the distal chromosome 2 imprinting region. Mouse Genome 1992; 90:82.
3. Peters J, Beechey CV, Ball ST et al. Mapping studies of the distal imprinting region of mouse chromosome 2. Genet Res 1994; 63:169-174.
4. Williamson CM, Schofield J, Dutton ER et al. Glomerular-specific imprinting of the mouse Gsα gene: how does this relate to hormone resistance in Albright's hereditary osteodystrophy. Genomics 1996; 36:280-287.
5. Davies SJ, Hughes HE. Imprinting in Albright's hereditary osteodystrophy. J Med Genet 1993; 30:101-103.
6. Yu S, Yu D, Lee E et al. Variable and tissue-specific hormone resistance in heterotrimeric G_s protein α-subunit ($G_s\alpha$) knockout mice is due to tissue-specific imprinting of the $G_s\alpha$ gene. Proc Nat Acad Sci USA 1998; 95:8715-8720.
7. Peters J, Holmes R, Monk D et al. Imprinting control within the compact Gnas locus. Cytogenet Genome Res 2006; 113:194-201.
8. Bastepe M, Fröhlich LF, Hendy GN et al. Autosomal dominant pseudohypoparathyroidism type 1b is associated with a heterozygous microdeletion that likely disrupts a putative imprinting control element of GNAS. J Clin Invest 2003; 112:1255-1263.
9. Linglart A, Gensure RC, Olney RC et al. A novel STX16 deletion in autosomal dominant pseudohypoparathyroidism type 1b redefines the boundaries of a cis-acting imprinting control element of GNAS. Am J Hum Genet 2005; 76:804-814.
10. Peters J, Wroe SF, Wells CA et al. A cluster of oppositely imprinted transcripts at the Gnas locus in the distal imprinting region of mouse chromosome 2. Proc Nat Acad Sci USA 1999; 96:3830-3835.
11. Kelsey G, Bodle D, Miller HJ et al. Identification of imprinted loci by methylation-sensitive representational difference analysis: Application to mouse distal Chromosome 2. Genomics 1999; 62:129-138.
12. Liu J, Yu S, Litman D et al. Identification of a methylation imprint mark within the mouse Gnas locus. Mol Cell Biol 2000; 20:5808-5817.
13. Williamson CM, Ball ST, Nottingham WT et al. A cis-acting control region is required exclusively for the tissue-specific imprinting of Gnas. Nat Genet 2004; 36:894-899.
14. Holmes R, Williamson C, Peters J et al. A comprehensive transcript map of the mouse Gnas imprinted complex. Genome Res 2003; 13:1410-1415.
15. Wroe SF, Kelsey G, Skinner JA et al. An imprinted transcript, antisense to Nesp, adds complexity to the cluster of imprinted genes at the mouse Gnas locus. Proc Nat Acad Sci USA 2000; 97:3342-3346.
16. Li T, Vu TH, Zeng Z-L et al. Tissue-specific expression of antisense and sense transcripts at the imprinted Gnas locus. Genomics 2000; 69:295-304.
17. Williamson CM, Skinner JA, Kelsey G et al. Alternative noncoding splice variants of Nespas, an imprinted gene antisense to Nesp in the Gnas imprinting cluster. Mamm Genome 2002; 13:74-79.
18. Abramowitz J, Grenet D, Birnbaumer M et al. XLαs, the extra-long form of the α-subunit of the Gs G protein, is significantly longer than suspected and so is its companion Alex. Proc Nat Acad Sci USA 2004; 101:8366-8371.
19. Royo H, Bortolin M-L, Seitz H et al. Small noncoding RNAs and genomic imprinting. Cytogenet Genome Res 2006; 113:99-108.
20. Coombes C, Arnaud P, Gordon E et al. Epigenetic properties and identification of an imprint mark in the Nesp-Gnasxl domain of the mouse Gnas imprinted locus. Mol Cell Biol 2003; 23:5475-5488.
21. Crawford AJ, Mutchler KJ, Sullivan BE et al. Neural expression of a novel alternatively spliced and polyadenylated Gsα transcript. J Biol Chem 1993; 268:9879-9885.
22. Kehlenbach RH, Matthey J, Huttner WB. XLαs is a new type of G protein. Nature 1994; 372:804-809.
23. Klemke M, Pasolli HA, Kehlenbach RH et al. Characterization of the extra-large G protein alpha-subunit XLαs II. Signal transduction properties. J Biol Chem 2000; 275:33633-33640.
24. Bastepe M, Gunes Y, Perez-Villamil B et al. Receptor-mediated adenylyl cyclase activation through XLαs, the extra-large variant of the stimulatory G protein α-subunit. Mol Endocrinol 2002; 16:1912-1919.
25. Pasolli HA, Klemke M, Kehlenbach RH et al. Characterization of the extra-large G protein α-subunit XLαs. I. Tissue distribution and subcellular localization. J Biol Chem 2000; 275:33622-33632.
26. Plagge A, Gordon E, Dean W et al. The imprinted signaling protein XLαs is required for postnatal adaptation to feeding. Nat Genet 2004; 36:818-826.
27. Klemke M, Kehlenbach RH, Huttner WB. Two overlapping reading frames in a single exon encode interacting proteins—a novel way of gene usage. EMBO J 2001; 20:3849-3860.
28. Ischia R, Lovisetti-Scamihorn P, Hogue-Angeletti R et al. Molecular cloning and characterisation of NESP55, a novel chromogranin-like precursor of a peptide with 5-HT_{1B} receptor antagonist activity. J Biol Chem 1997; 272:11657-11662.

29. Plagge A, Isles AR, Gordon E et al. Imprinted Nesp55 influences behavioral reactivity to novel environments. Mol Cell Biol 2005; 25:3019-3026.
30. Bauer R, Weiss C, Marksteiner J et al. The new chromogranin-like protein NESP55 is preferentially localized in adrenaline-synthesizing cells of the bovine and renal medulla. Neurosci Lett 1999; 263:13-16.
31. Williamson CM, Turner MD, Ball ST et al. Identification of an imprinting control region affecting the expression of all transcripts in the Gnas cluster. Nat Genet 2006; 38:350-355.
32. Cattanach BM, Peters J, Ball S et al. Two imprinted gene mutations: three phenotypes. Hum Mol Genet 2000; 9:2263-2273.
33. Chen M, Gavrilova O, Liu J et al. Alternative Gnas gene products have opposite effects on glucose and lipid metabolism. Proc Nat Acad Sci USA 2005; 102:7386-7391.
34. Germain-Lee EL, Schwindinger W, Crane JL et al. A mouse model of Albright hereditary osteodystrophy generated by targeted disruption of exon 1 of the Gnas gene. Endocrinology 2005; 146:4697-4709.
35. Yu S, Gavrilova O, Chen H et al. Paternal versus maternal transmission of a stimulator G-protein α-subunit knockout produces opposite effects on energy metabolism. J Clin Invest 2000; 105:615-623.
36. Skinner JA, Cattanach BM, Peters J. The imprinted oedematous-small mutation on mouse chromosome 2 identifies new roles for Gnas and Gnasxl in development. Genomics 2002; 80:373-375.
37. Xie T, Plagge A, Gavrilova O et al. The alternative stimulatory G protein α-subunit XLαs is a critical regulator of energy and glucose metabolism and sympathetic nerve activity in adult mice. J Biol Chem 2006; 281:18989-18999.
38. Wilkins JF, Haig D. What good is genomic imprinting: the function of parent-specific gene expression. Nature 2003; 4:1-10.
39. Lewis A, Reik W. How imprinting centres work. Cytogenet Genome Res 2006: 113:81-89.
40. Spahn L, Barlow DP. An ICE pattern crystallizes. Nat Genet 2003; 35(1):11-12.
41. Sakamoto A, Liu J, Greene A et al. Tissue-specific imprinting of the G protein $G_s\alpha$ is associated with tissue-specific differences in histone methylation. Hum Mol Genet 2004; 13:819-828.
42. Choo JJ, Kim JD, Chung JH et al. Allele-specific deposition of macroH2A1 in imprinting control regions. Hum Mol Genet 2006; 15:717-724.
43. Kim JD, Hinz AK, Bergmann A et al. Identification of clustered YY1 binding sites in imprinting control regions. Genome Res 2006; 16:901-911.
44. Bastepe M, Fröhlich LF, Linglart A et al. Deletion of the NESP55 differentially methylated region causes loss of maternal GNAS imprints and pseudohypoparathyroidism type Ib. Nat Genet 2005; 37:25-27.
45. Bastepe M. (This volume)
46. Liu J, Nealon JG, Weinstein LS. Distinct patterns of abnormal GNAS imprinting in familial and sporadic pseudohypoparathyroidism type 1B. Hum Mol Genet 2005; 14:95-102.
47. Liu J, Chen M, Deng C et al. Identification of the control region for tissue-specific imprinting of the stimulatory G protein α-subunit. Proc Nat Acad Sci USA 2005; 102:5513-5518.
48. Williamson CM, Turner MD, Ball ST et al. Identification of an imprinting control region affecting the expression of all transcripts in the Gnas cluster. Nat Genet 2006; 38:350-355.
49. Lopes S, Lewis A, Hajkova P et al. Epigenetic modifications in an imprinting cluster are controlled by a hierarchy of DMRs suggesting long-range chromatin interactions. Hum Mol Genet 2003; 12:295-305.
50. Murrell A, Heeson S, Reik W. Interaction between differentially methylated regions partitions the imprinted genes Igf2 and H19 into parent-specific chromatin loops. Nat Genet 2004; 36:889-893.
51. Lyle R, Watanabe D, Vruchte D et al. The imprinted antisense RNA at the Igf2r locus overlaps but does not imprint Mas1. Nat Genet 2000; 25:19-21.
52. Mitsuya K, Meguro M, Lee MP et al. LIT1, an imprinted antisense RNA in the human KvLQT1 locus identified by screening for differentially expressed transcripts using monochromosomal hybrids. Hum Mol Genet 1999; 8:1209-1217.
53. Rougeulle C, Cardoso C, Fontés M et al. An imprinted antisense RNA overlaps UBE3A and a second maternally expressed transcript. Nat Genet 1998; 19:15-16.
54. Wutz A, Smrzka OW, Schweifer N et al. Imprinted expression of the Igf2r gene depends on an intronic CpG island. Nature 1997; 389:745-749.
55. Fitzpatrick GV, Soloway PD, Higgins MJ. Regional loss of imprinting and growth deficiency in mice with a targeted deletion of KvDMR1. Nat Genet 2002; 32:426-431.
56. Sleutels F, Zwart R, Barlow DP. The noncoding Air RNA is required for silencing autosomal imprinted genes. Nature 2002; 415:810-813.
57. Mancini-DiNardo D, Steele SJS, Levorse JM et al. Elongation of the Kcnq1ot1 transcript is required for genomic imprinting on neighboring genes. Genes Dev 2006; 20:1268-1282.
58. Weinstein LS, Yu S, Warner DR et al. Endocrine manifestations of stimulatory G protein α-subunit mutations and the role of genomic imprinting. Endocrine Reviews 2001; 22:675-705.

CHAPTER 3

The *GNAS* Locus and Pseudohypoparathyroidism

Murat Bastepe*

Abstract

Pseudohypoparathyroidism (PHP) is a disorder of end-organ resistance primarily affecting the actions of parathyroid hormone (PTH). Genetic defects associated with different forms of PHP involve the α-subunit of the stimulatory G protein (Gsα), a signaling protein essential for the actions of PTH and many other hormones. Heterozygous inactivating mutations within Gsα-encoding *GNAS* exons are found in patients with PHP-Ia, who also show resistance to other hormones and a constellation of physical features called Albright's hereditary osteodystrophy (AHO). Patients who exhibit AHO features without evidence for hormone resistance, who are said to have pseudopseudohypoparathyroidism (PPHP), also carry heterozygous inactivating Gsα mutations. Maternal inheritance of such a mutation leads to PHP-Ia, i.e., AHO *plus* hormone resistance, while paternal inheritance of the same mutation leads to PPHP, i.e., AHO only. This imprinted mode of inheritance for hormone resistance can be explained by the predominantly maternal expression of Gsα in certain tissues, including renal proximal tubules. Patients with PHP-Ib lack coding Gsα mutations but display epigenetic defects of the *GNAS* locus, with the most consistent defect being a loss of imprinting at the exon A/B differentially methylated region (DMR). This epigenetic defect presumably silences, *in cis*, Gsα expression in tissues where this protein is derived from the maternal allele only, leading to a marked reduction of Gsα levels. The familial form of PHP-Ib (AD-PHP-Ib) is typically associated with an isolated loss of imprinting at the exon A/B DMR. A unique 3-kb microdeletion that disrupts the neighboring *STX16* locus has been identified in this disorder and appears to be the cause of the loss of imprinting. In addition, deletions removing the entire NESP55 DMR, located within *GNAS*, have been identified in some AD-PHP-Ib kindreds in whom affected individuals show loss of all the maternal *GNAS* imprints. Mutations identified in different forms of PHP-Ib thus point to different *cis*-acting elements that are apparently required for the proper imprinting of the *GNAS* locus. Most sporadic PHP-Ib cases also have imprinting abnormalities of *GNAS* that involve multiple DMRs, but the genetic lesion(s) responsible for these imprinting abnormalities remain to be discovered.

Introduction

Many hormones, neurotransmitters and autocrine/paracrine factors exert their actions through receptors coupled to Gsα, one of the several gene products of the imprinted *GNAS* locus (see the chapter by Peters and Williamson). Gsα, when activated by an agonist-occupied G protein-coupled receptor, stimulates adenylyl cyclase, thereby generating the second messenger cyclic AMP. Demonstrating the reliance of many developmental processes to Gsα, homozygous disruption of

*Murat Bastepe—Endocrine Unit, Department of Medicine, Massachusetts General Hospital and Harvard Medical School, Boston, MA 02114, U.S.A.
Email: bastepe@helix.mgh.harvard.edu

Genomic Imprinting, edited by Jon F. Wilkins. ©2008 Landes Bioscience and Springer Science+Business Media.

this signaling protein in the mouse is lethal during early embryonic development.[1-3] Furthermore, genetic defects affecting even a single *GNAS* allele are associated with human disease. Somatic mutations that constitutively activate Gsα are found in various endocrine tumors, such as growth hormone secreting adenomas and postzygotic activating mutations are found in patients with McCune-Albright syndrome (reviewed in refs. 4, 5). Heterozygous mutations within *GNAS* that impair either the activity or the expression of Gsα are associated with pseudohypoparathyroidism (PHP), a disorder of target-organ resistance affecting predominantly, but not exclusively, the actions of parathyroid hormone (PTH). While the imprinting of the *GNAS* locus is predicted to influence the molecular mechanisms in all of these disorders, its role has been best documented in the development of PHP, which includes various different clinical types that are caused by related, but distinct, genetic defects and show parent-of-origin-specific inheritance.

PTH is the primary regulator of serum calcium, acting mainly on kidney and bone as target organs via its primarily Gsα-coupled receptor PTHR1 (reviewed in refs. 6, 7). Secretion of PTH from the parathyroid gland is tightly regulated and increases in response to low serum calcium. In the renal proximal tubule, PTH increases the mRNA level of 25-hydroxyvitamin D 1-α hydroxylase. This leads to an elevation of serum 1,25 dihydroxyvitamin D3, the active vitamin D metabolite, which in turn enhances the intestinal absorption of calcium and phosphate. PTH also negatively regulates the expression and subcellular distribution of the type-IIa sodium-phosphate cotransporter in the renal proximal tubule and, thereby, inhibits reabsorption of phosphate. In the distal nephron, PTH improves reabsorption of calcium that occurs through transcellular mechanisms. The physiologic action of PTH in the bone is mainly resorptive, leading to mobilization of both calcium and phosphate from bone into the circulation. In PHP, resistance to PTH appears to occur only in the renal proximal tubule, whereas the actions of PTH are apparently unimpaired in other target tissues, such as bone[8,9] and the thick ascending tubule.[10] Hence, patients with PHP have reduced serum concentrations of 1,25-dihydroxy vitamin D3[11,12] and hypocalcemia. Serum phosphate level is typically elevated due to the inability of PTH to inhibit phosphate reabsorption in the proximal tubule and due to the unimpaired resorptive action of PTH on bone that leads to the mobilization of phosphate (and calcium). Illustrating the PTH-resistance in the renal proximal tubule, administration of exogenous biologically active PTH fails to result in an appropriate increase in urinary phosphate and, in certain forms (see below), urinary cAMP.[13,14] As expected, serum PTH concentration is elevated in patients with PHP, indicating that target-organ resistance rather than deficiency of PTH (hypoparathyroidism) is the underlying defect. Because of the unimpaired PTH functions, elevated PTH concentration can maintain normal serum calcium level in some PHP patients for prolonged periods of time. However, most of these patients develop, at some point of their lives, hypocalcemia with associated clinical manifestations, such as muscle spasms or seizures and require treatment with oral calcium supplements and 1,25 dihydroxyvitamin D preparations. In addition, those asymptomatic patients with normal serum calcium and phosphate but with elevated serum PTH should also be treated in order to normalize the PTH level and, thereby, prevent bone resorption, which could lead to hyperparathyroid bone disease.[15] Patients undergoing treatment should be monitored annually for both blood biochemistries and urinary calcium excretion to avoid persistent hypercalcemia and/or hypercalciuria.[16]

Since its first description in 1942 by Albright and colleagues,[14] PHP has been subdivided into two major clinical types depending on whether urinary excretion of both cAMP and phosphate are blunted (type I) or urinary excretion of only phosphate is blunted (type-II) following exogenous administration of PTH as a diagnostic test.[13,14] Thus far, only few cases of PHP-II have been reported and the nature of the molecular defect responsible for this PHP variant remains elusive. On the other hand, PHP-I is relatively common and various molecular defects underlying this form of PHP have been identified. This chapter will therefore focus on PHP-I and its subtypes, although it should be noted that the pharmacological management of the clinical findings that result from hormone resistance in PHP-Ia remains currently the same as those in PHP-II.

Inactivating Gsα Mutations and Multiple Hormone Resistance: PHP-Ia

Patients with PHP-I are subdivided into PHP-Ia and PHP-Ib, depending on the presence or absence of additional hormone resistance and Albright's hereditary osteodystrophy (AHO), a constellation of physical features, including obesity, short stature, ectopic ossification in subcutaneous tissues, brachydactyly and/or mental retardation (Table 1). In addition to the clinical signs of PTH-resistance, nearly all patients with PHP-Ia exhibit mild hypothyroidism due to resistance to thyroid stimulating hormone (TSH).[17,18] In addition, hypogonadism and growth hormone deficiency can also be present in PHP-Ia, reflecting resistance to the actions of gonadotropins and growth hormone releasing hormone (GHRH), respectively.[17,19-21] In contrast, PHP-Ia patients apparently show intact responses to many other hormones that signal through Gsα, such as vasopressin[22,23] and those in the hypothalomo-pituitary-adrenal axis.[20,22,24]

Hormone resistance in patients with PHP-Ia results from heterozygous inactivating mutations that affect *GNAS* exons encoding Gsα.[25,26] This finding correlates well with the observation that all the hormones whose actions are impaired in PHP-Ia mediate their actions primarily through Gsα-coupled receptors. Inactivating mutations in PHP-Ia patients have been identified in nearly all of the thirteen exons that encode the Gsα protein and include missense and nonsense amino acid changes, as well as insertions and deletions that either alter pre-mRNA splicing or introduce frame-shifts causing early termination. Identified mutations also include constitutional deletions of chromosome 20q,[27] the chromosomal region that comprises the *GNAS* locus (a list of mutations associated with PHP-Ia and AHO can be found under OMIM entry #139320 at http://www.ncbi.nlm.nih.gov). Most of the Gsα mutants are not properly expressed either due to mRNA instability or due to altered subcellular localization. Accordingly, accessible tissues from PHP-Ia patients, such as skin fibroblasts and erythrocytes, reveal a ~50% reduction in Gsα mRNA/protein.[28-30] In addition, biochemical assays involving reconstitution of patient-derived cell membranes with membranes of cells that lack functional Gsα show an approximately 50% reduction in hormone-induced cAMP generation.[31,32] If non-hydrolysable GTP analogs are used as stimulants to assess Gsα activity, however, these assays show 100% functionality for Gsα mutants that are

Table 1. *Clinical features and molecular defects associated with different subtypes of PHP-I*

	PTH Resistance	Other Hormone Resistance[a]	Typical AHO Features[b]	Genetic Lesion	GNAS Epigenetic Defect	Parental Inheritance
PHP-Ia	Yes	Yes	Yes	Loss-of-function Gsα mutations	No	Maternal
PPHP	No	No	Yes	Loss-of-function Gsα mutations	No	Paternal
POH	No	No	No	Loss-of-function Gsα mutations	No	Paternal[c]
PHP-Ib	Yes	TSH-resistance in some cases	No	Deletions involving STX16 or NESP55	Yes	Maternal

[a]Other hormones with impaired actions typically include TSH, gonadotropins and GHRH.
[b]AHO features include short-stature, obesity, brachydactyly, ectopic ossification and mental deficits.
[c]Heterotopic ossifications are rarely seen in patients who inherit Gsα mutations maternally; however, these patients also show hormone resistance and, mostly, some AHO features.

defective in receptor coupling, but not in adenylyl cyclase stimulation.[33,34] The latter observation reflects primarily the response to the type of stimulant used in the functional assay rather than the clinical phenotype, but it fits with the definition of another PHP-I subtype, termed PHP-Ic, which is used to describe patients who have the clinical characteristics of PHP-Ia but display normal Gsα bioactivity.[35] It remains to be determined whether the so-called PHP-Ic patients indeed represent a subgroup of PHP-Ia patients that carry mutations within the receptor coupling Gsα domains, or if they constitute a distinct group in whom the genetic defect lies downstream of the receptor activated cAMP generation.

PHP-Ia is closely related to two disorders at both the molecular and clinical levels (Table 1): pseudopseudohypoparathyroidism (PPHP) and progressive osseous heteroplasia (POH). PPHP is a term coined by Albright and colleagues[36] in order to describe patients who present with the typical features of AHO but lack any evidence for hormone resistance. Patients with PPHP also carry heterozygous inactivating mutations in Gsα-coding *GNAS* exons and in fact, PPHP and PHP-Ia occur within the same kindreds.[25,37] A careful analysis of several such kindreds has revealed that the inheritance of each disorder follows an imprinted mode, i.e., the phenotype of the offspring is determined by the gender of the parent transmitting the molecular defect rather than his/her phenotype. According to this imprinted mode of inheritance, the genetic defect leads to AHO without hormone resistance (i.e., PPHP) upon inheritance from a male patient with either PHP-Ia or PPHP, whereas it leads to both AHO and hormone resistance (i.e., PHP-Ia) upon inheritance from a female patient with either disorder.[38,39] In other words, hormone resistance develops only when the Gsα mutation is inherited maternally, while AHO develops when the Gsα mutation is inherited from either parent. This imprinted mode of inheritance for hormone resistance is consistent with the known imprinting of the *GNAS* locus and the evidence that the protein product Gsα shows predominantly maternal expression in certain tissues.

POH is a disorder of severe heterotopic ossifications that involve not only the subcutaneous tissues, as seen typically in PHP-Ia, but also the skeletal muscle and deep connective tissue, often leading to severe malformation of neighboring tissues.[40] Patients with POH have also been found to carry heterozygous mutations in the *GNAS* exons encoding Gsα and some of the mutations associated with POH are identical to those found in patients with either PHP-Ia or PPHP.[40-42] Furthermore, some patients exhibit POH in combination with hormone resistance and typical AHO features,[41,43] suggesting that at least some cases of POH may be an extreme manifestation of the heterotopic ossification seen in AHO. On the other hand, the POH phenotype in most reported cases is manifest as an isolated finding and develops only after paternal inheritance of a Gsα mutation,[42] suggesting that additional mechanisms that involve imprinting and/or additional genetic modifiers take part in the pathogenesis of this disorder.

Role of Tissue- and Parental Origin-Specific Gsα Expression in Hormone Resistance

Despite the importance of Gsα signaling in the actions of many different hormones, PHP-Ia patients show resistance only to a limited number of hormones acting through Gsα-coupled receptors. Furthermore, this hormone resistance becomes manifest only after maternal transmission of a Gsα mutation. These findings have suggested a disease mechanism that involves monoallelic expression of Gsα from the maternally derived chromosome in some but not all tissues. In tissues where Gsα is expressed only or predominantly from the maternal *GNAS* allele, such as the renal proximal tubule,[1] a Gsα mutation causes nearly complete loss of Gsα expression when inherited maternally, therefore leading to hormone resistance. The same Gsα mutation does not affect the Gsα expression in those tissues when inherited paternally and thus, the hormone responses remain intact. On the other hand, in tissues where Gsα is biallelic, a Gsα mutation results in an approximately 50% reduction regardless of the parental origin. In most tissues, this 50% reduction appears to be sufficient for maintaining normal signaling activity, such as in renal medulla, where the actions of vasopression appear to be intact.[22,23] In some other tissues, however, the levels of Gsα is more critical and the 50% reduction in Gsα levels results in haploinsufficiency;[44] thus, AHO features develop regardless of the gender of the parent transmitting the Gsα defect.

The tissue-specific, monoallelic expression of Gsα has been demonstrated originally upon generation of a mouse model by Yu et al[1] who knocked out this protein through targeted disruption of *Gnas* exon 2; at the time, this exon was considered to be unique to Gsα, but it is now known to be shared by other *GNAS* transcripts (see the chapter by Peters and Williamson). Based on the analysis of this mouse strain, homozygous ablation of Gsα leads to embryonic lethality during early postimplantation, a finding that correlates well with the requirement of Gsα in numerous physiological processes. Consistent with the findings in kindreds with PHP-Ia/PPHP, mice with maternal (*Gnas*$^{E2+/mat-}$) or paternal (*Gnas*$^{E2+/pat-}$) disruption of *Gnas* exon 2 that survive beyond weaning are significantly smaller than their wild-type littermates.[1] Furthermore, *Gnas*$^{E2+/mat-}$ mice show resistance to PTH in the proximal tubule, as evidenced by a marked reduction in the amount of PTH-induced cAMP production in proximal tubules isolated from the kidneys of these mice. In addition, the *Gnas*$^{E2+/mat-}$ mice develop hypocalcemia and hyperphosphatemia and show elevated serum PTH levels. In contrast, *Gnas*$^{E2+/pat-}$ mice show normal responsiveness to PTH in the renal proximal tubule and, accordingly, show no evidence for defective calcium metabolism. Also consistent with findings in PHP-Ia/PPHP kindreds, hormone responsiveness of renal medulla appears intact regardless of the parental origin of *Gnas* exon 2 disruption. Protein and mRNA expression analysis indicates that *Gnas*$^{E2+/mat-}$ mice almost completely lack Gsα expression in the renal cortex, whereas *Gnas*$^{E2+/pat-}$ mice have Gsα levels indistinguishable from those of wild-type mice in this portion of the kidney. On the other hand, Gsα levels in renal medulla is reduced by about 50% in both *Gnas*$^{E2+/mat-}$ and *Gnas*$^{E2+/pat-}$ mice. Thus, this pattern of Gsα expression in the kidneys of *Gnas*$^{E2+/mat-}$ and *Gnas*$^{E2+/pat-}$ mice is concordant with the observed phenotypes with respect to hormone responsiveness. Moreover, these findings demonstrate that while expression of Gsα is biallelic in renal medulla, it is monoallelic and almost completely maternal in renal cortex.

Predominantly maternal Gsα expression has been shown in some other tissues, such as thyroid, through the study of another *Gnas* knockout mouse strain generated through disruption of exon 1.[3] Monoallelic, maternal-specific Gsα expression in the thyroid is consistent with the TSH-resistance observed in PHP-Ia patients. *Gnas* exon 1 knockout mice have been generated independently by two separate groups of investigators,[2,3] and it appears that this mouse strain recapitulates the clinical findings observed in PHP-Ia, including obesity and short-stature, to a greater extent than does the *Gnas* exon 2 knockout mouse strain. While this is consistent with the preservation of the additional *Gnas* transcripts that also use exon 2 (see below) in the *Gnas* exon 1 knockout strain, there are significant differences in the degree of obesity between mice that inherit the disrupted exon 1 allele maternally and those that inherit the disrupted allele paternally. Since Gsα appears to be biallelic in the white adipose tissue,[2,3] these differences could reflect disruption of as-yet-undefined *Gnas* transcripts that also use exon 1 and show parent-of-origin-specific expression. Alternatively, the difference in the degree of obesity may be due to systemic effects on the fat tissue caused by disrupted expression of Gsα in tissues where this protein is normally monoallelic. For a further discussion of the metabolic consequences of mutations at the *Gnas* locus, see the chapter by Frontera et al.

Gsα expression also appears to be biallelic in the growth plate.[44] As shown in mice carrying a conditional knockout of Gsα in this tissue[45] and mice chimeric for wild-type cells and cells homozygous for disruption of *Gnas* exon 2,[44] Gsα plays an essential role in PTH-related protein (PTHrP)-mediated delay of chondrocyte differentiation. Although no detectable abnormalities are present in the growth plates of mice with heterozygous disruption of *Gnas* exon 2,[1] a modest acceleration in the hypertrophic differentiation of growth plate chondrocytes has been revealed in chimeric mice that contain cells heterozygous for disruption of *Gnas* exon 2.[44] This finding is consistent with the notion that the short-stature and/or brachydactyly seen in AHO results, at least in part, from Gsα haploinsufficiency in the growth plate.

The tissue profile of monoallelic Gsα expression in humans appears to be largely similar to that in mice. Recent investigations of adult human tissues have shown predominant maternal expression of Gsα in thyroid gland,[46-48] ovaries,[46] and pituitary.[49] Biallelic Gsα expression has been documented in several other tissues, including adrenal gland, bone and adipose tissue.[46,50] Parental origin of Gsα expression has yet to be examined in human renal proximal tubules. However, analysis of human fetal kidney cortex using RT-PCR has demonstrated biallelic Gsα expression.[51] Since the maternal-specific inheritance of PTH-resistance in PHP-Ia patients strongly suggests predominantly maternal Gsα expression in the renal proximal tubule, the latter finding could suggest that Gsα imprinting takes place only in a small subset of renal cortical cells or that it establishes later in life in this tissue. The latter hypothesis is consistent with the finding that PTH-resistance in patients with PHP-Ia and PHP-Ib (see below) is typically not present at birth but rather develops after infancy.[52-54]

Overall, the results of these investigations show that Gsα expression is monoallelic in certain tissues. This parent-of-origin-specific expression profile of Gsα correlates well with the finding in PHP-Ia/PPHP kindreds that hormone resistance develops only after maternal inheritance of a Gsα mutation. Furthermore, the finding that the monoallelic expression of Gsα occurs only in a limited number of tissues is consistent with the observation that PHP-Ia patients exhibit resistance to only a limited number of hormones that act through Gsα.

Mutations Affecting the Imprinting Control Regions of *GNAS* and PTH-Resistance: PHP-Ib

Some patients with PHP present with PTH-resistance but lack any AHO features, defining another subtype of PHP termed PHP-Ib. As seen in PHP-Ia, the severity of PTH resistance in PHP-Ib varies significantly from patient to patient.[55,56] Furthermore, some patients with PHP-Ib can present with additional mild TSH-resistance.[48,56,57] Although most PHP-Ib cases are sporadic, i.e., there are no other family members known to be affected with the same disorder, a significant fraction of cases are familial. In the latter cases, the disease appears to show autosomal dominant inheritance with incomplete penetrance (AD-PHP-Ib), but in fact, it follows a parent-of-origin-specific mode of inheritance (see below). Patients with PHP-Ib typically lack mutations within the *GNAS* exons encoding Gsα.[17,58] Therefore, PHP-Ib was originally thought to be caused by mutations at an entirely different locus. Mutations in the most likely candidate, the gene encoding PTHR1, were ruled out in a number of patients by several independent investigations.[59-63] A genome-wide genetic linkage analysis in four large AD-PHP-Ib kindreds has mapped the disease gene to a region on chromosome 20q that contains *GNAS* at its telomeric boundary.[55] This location has been confirmed by subsequent studies,[56,64] which excluded the coding regions of the *GNAS* locus from the critical region.[56] Furthermore, analysis of large PHP-Ib kindreds have revealed that the PTH-resistance develops only following inheritance of the genetic defect from a female obligate gene carrier[55] (Fig. 1A), i.e., the inheritance mode of hormone resistance in PHP-Ib is identical to that seen in PHP-Ia/PPHP kindreds. In addition, most sporadic PHP-Ib cases and patients with AD-PHP-Ib exhibit alterations in the imprinting of the *GNAS* locus (see the chapter by Peters and Williamson for a complete discussion pertaining to the imprinting of this locus). Although such abnormalities can be found in various *GNAS* DMRs in different cases, the most consistent imprinting defect found in PHP-Ib patients is a loss of methylation at the exon A/B DMR, combined with biallelic expression of the A/B transcript[65] (Fig. 1B). (Note that while the mouse homolog of this transcript is called 1A,[66] this transcript has been originally called A/B in humans.[67]) Loss of exon A/B imprinting is frequently found as an isolated defect in AD-PHP-Ib kindreds.[56] Based on these findings, it appears that PHP-Ib is caused by mutations that disrupt *cis*-acting elements regulating imprinting of *GNAS*.

Consistent with hormone resistance being the result of disturbed imprinting, a sporadic PHP-Ib case was shown to have a paternal uniparental isodisomy of the entire long arm of chromosome 20.[57] A defect in Gsα activity in tissues with biallelic expression was ruled out in this PHP-Ib patient by demonstration of normal Gsα level and activity in skin fibroblasts.[57] In contrast to typical PHP-Ib patients, this patient also exhibited developmental delay and craniosynostosis, which may have resulted from unmasking of recessive defects on the duplicated paternal chromosome

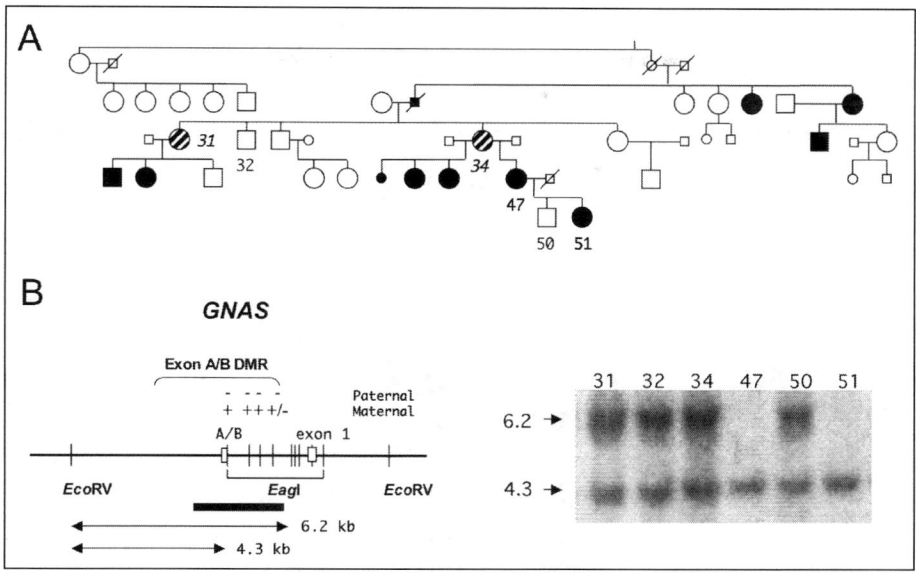

Figure 1. PHP-Ib patients demonstrate loss of GNAS imprinting. Molecular studies have revealed epigenetic alterations within GNAS in both sporadic and familial cases of PHP-Ib. The most consistent change, however, is a loss of imprinting at the exon A/B DMR, which is observed as an isolated defect in most AD-PHP-Ib kindreds. A) The pedigree of an AD-PHP-Ib kindred. Black symbols, affected; white symbols, unaffected; striped symbols, obligate gene carriers. Note that all affected individuals have inherited the disease from their mothers. B) Southern blot analysis of differential methylation at exon A/B, performed through the use of methylation-sensitive (EagI) and –insensitive (EcoRV) restriction enzymes. Exons and introns are depicted by boxes and connecting lines, respectively. Methylated (+) and nonmethylated (–) alleles, as well as restriction fragment lengths relevant to the Southern blot analysis are indicated. Horizontal bar, Southern probe.

20q and/or abnormalities due to disrupted expression of other gene products of GNAS, such as XLαs. In addition, disrupted gene expression from other imprinted loci located on 20q, such as NNAT,[68-70] might also account for the unique phenotypic features in this patient. Nonetheless, this case demonstrates the importance of imprinting in the pathogenesis of PHP-Ib and shows that a paternal uniparental disomy that involves the GNAS locus can lead to hormone resistance.

In affected individuals and unaffected carriers of multiple unrelated AD-PHP-Ib kindreds, genetic linkage and nucleotide sequence analyses of the linked region have revealed a heterozygous 3-kb microdeletion[71] (Fig. 2). The same mutation has been subsequently found in numerous other AD-PHP-Ib kindreds;[72-74] this mutation is thus apparently a frequent cause of this disorder. Correlating well with the mode of inheritance in AD-PHP-Ib, PTH-resistance is associated with maternal inheritance of this deletion in each of these kindreds. Flanking the identified 3-kb microdeletion are two direct repeats of 391 bp, suggesting that the mutation is mediated by homologous recombination between these two repeats. Such a mechanism, which would cause deletions at a higher frequency than most other mutations, is indeed consistent with the finding that the same mutation is found in multiple kindreds of different racial and ethnic origin.

The 3-kb microdeletion is located approximately 220 kb upstream of exon A/B and removes exons 4-6 of another gene, STX16, which encodes a member of the SNARE (soluble N-ethylmaleimide-sensitive factor attachment protein receptor) family of proteins involved in intracellular trafficking and vesicle fusion[75,76] (Fig. 2). Recently, a different deletion (4.4-kb) has been discovered in a different AD-PHP-Ib kindred, causing PTH-resistance only after its transmission from

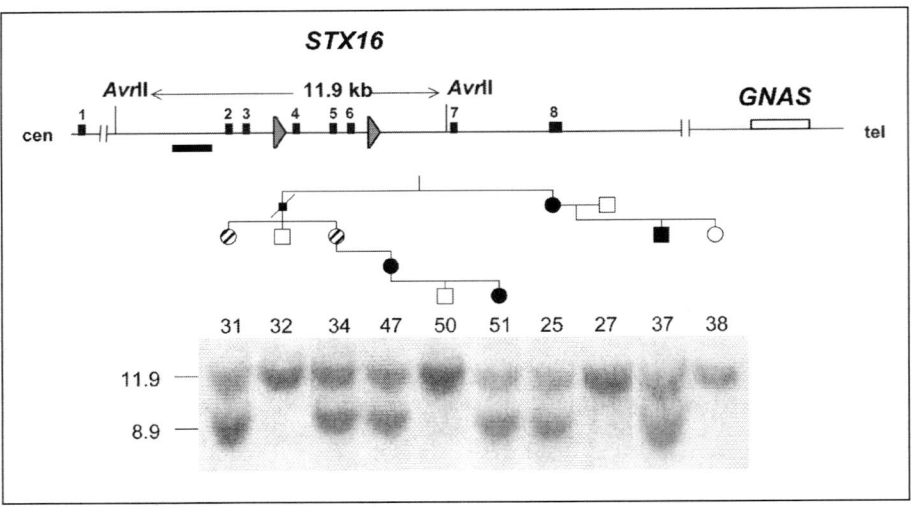

Figure 2. Maternal inheritance of a unique 3-kb microdeletion within *STX16* is a frequent cause of AD-PHP-Ib. The deleted region removes *STX16* exons 4-6 and is flanked by two direct repeats (triangles) of 391 bp. *GNAS* exon A/B is located ~220 kb downstream of this region. Exons and introns are depicted as black boxes and connecting lines, respectively. Horizontal bar, Southern probe; white rectangle, the entire *GNAS* locus. Southern blot analysis reveals both the wild-type allele (11.9 kb) and the mutant allele (8.9 kb) in genomic DNA from affected members and unaffected carriers of an AD-PHP-Ib kindred.

female carriers.[54] This new deletion, which overlaps with the previously identified 3-kb deletion by approximately 1.3-kb, also disrupts the *STX16* locus by removing exons 2-4. Thus, both deletions are predicted to disrupt the syntaxin-16 transcript derived from the maternal chromosome. Nonetheless, given that AD-PHP-Ib shows an imprinted mode of inheritance, a loss of one copy of *STX16* could cause PTH-resistance only if this gene were imprinted. Yet, a number of findings argue against this possibility; (i) the promoter of *STX16* lack differential methylation,[71] (ii) both deleted and wild-type syntaxin-16 transcripts can be amplified from lymphoblastoid cells derived from patients,[54] and (iii) syntaxin-16 transcripts show biallelic expression in lymphoblastoid cells derived from normal individuals.[54] Although these findings do not rule out the possibility that *STX16*, like Gsα, shows parent-of-origin-specific expression in a tissue-specific manner, it appears unlikely that disruption of one *STX16* allele is responsible for the pathogenesis of AD-PHP-Ib. On the other hand, there is a perfect correlation between the identified *STX16* deletions, upon maternal inheritance and the nature of the *GNAS* epigenetic defects, i.e., a loss of exon A/B imprinting without changes at other *GNAS* DMRs.[54,71-74] Therefore, it appears likely that the *STX16* locus harbors a *cis*-acting control element required for the establishment and/or maintenance of the methylation imprint at exon A/B (Fig. 3); however, it remains conceivable that syntaxin-16 protein has a critical role in the establishment of exon A/B methylation in the female germ cell. When compared to the regions of synteny in mouse and rat, the *STX16* locus is well conserved with respect to intron-exon architecture. The only highly conserved nucleotide sequence within the deletion overlap corresponds to *STX16* exon 4, which lies within a small CpG-rich segment. This region may thus comprise the putative *cis*-acting element controlling the imprinting of *GNAS* exon A/B, although it lacks differential methylation itself.[71] The generation and study of mouse models carrying deletions at the mouse *STX16* locus will be required in order to confirm the role of this region in *GNAS* imprinting.

Most sporadic PHP-Ib and some AD-PHP-Ib cases display imprinting abnormalities that affect the entire *GNAS* locus,[56,65,74,77] and deletions within *STX16* could not be found in these cases.[71,74]

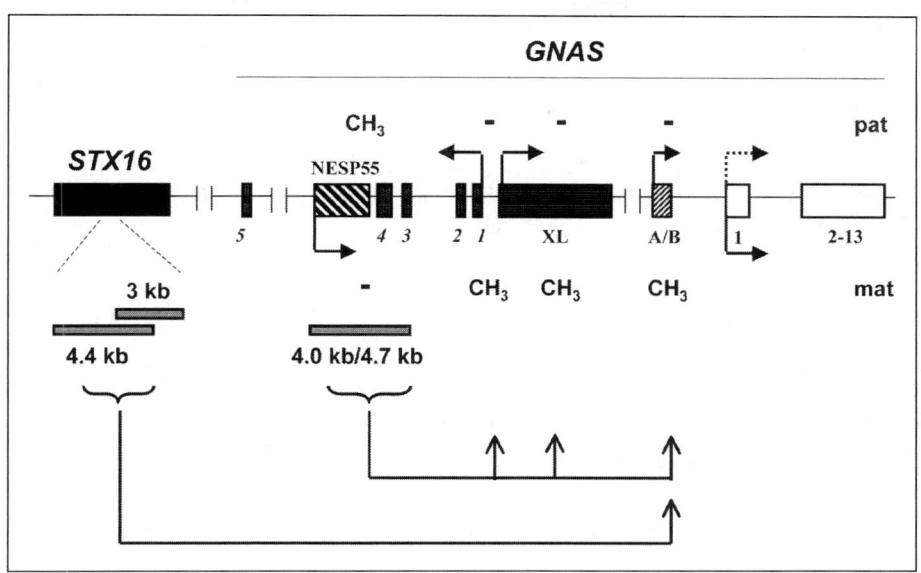

Figure 3. Mutations identified in patients with PHP-Ib reveal putative *cis*-acting elements regulating *GNAS* imprinting. Deletions within *STX16* appear to disrupt a *cis*-acting control element of *GNAS* that is required for the imprint mark located at exon A/B. Deletions of the NESP55 DMR in some AD-PHP-Ib kindreds reveal a cis-acting element controlling imprinting of the entire maternal *GNAS* allele. The epigenetic change at exon A/B and/or the associated derepression of maternal A/B transcript are postulated to silence, in *cis*, Gsα expression in those tissues in which this protein is normally silenced from the paternal *GNAS* allele, e.g., renal proximal tubules. This is predicted to result in a marked reduction of Gsα expression, leading to PTH-resistance. Boxes and connecting lines indicate exons and introns, respectively. *STX16* exons and *GNAS* exons 2-13 are shown as single rectangles for simplicity. Paternal (pat) and maternal (mat) methylation (CH₃) and parental origin of transcription (arrows) are marked. Dotted arrow indicates the tissue-specific silencing of the paternal Gsα transcription. Grey horizontal bars indicate the deletions identified in patients with AD-PHP-Ib.

In two AD-PHP-Ib kindreds displaying broad *GNAS* imprinting defects, large deletions that include the exon NESP55 DMR have been identified; these deletions also include exons 3 and 4 of the *GNAS* antisense transcript.[77] Of note, the deletion of the NESP55 DMR is accompanied, in one of these kindreds, by an approximately 15-kb insertion duplicated from the nearby region between *GNAS* exons XL and A/B. Due to a lack of a similar genomic rearrangement in the other AD-PHP-Ib kindred, however, it appears that the NESP55 deletion is the most likely cause of the imprinting abnormalities common to these two kindreds. Consistent with the parent-of-origin specific inheritance of PTH-resistance in AD-PHP-Ib, the identified NESP55 deletions are maternally inherited in affected individuals.[77] The maternal inheritance affects all the maternal *GNAS* methylation imprints and causes biallelic expression of the antisense transcript, XLαs and the A/B transcript. Molecular analysis of genomic DNA from patients carrying the maternally inherited NESP55 deletion shows an apparent gain of methylation of this region. Conversely, the unaffected carriers in these kindreds, who carry paternally inherited NESP55 deletions, display an apparent loss of methylation at the same site.[56,77] The epigenetic changes associated with the maternal NESP55 deletion suggest that this region harbors yet another *cis*-acting element required for the proper imprinting of the maternal *GNAS* allele (Fig. 3). Studies of the mouse *GNAS* cluster have shown that the NESP55 DMR is established during mid-gestation and, thus, do not represent a germ-line imprint.[66,78] Therefore, if methylation-sensitive binding of a trans-acting factor were required for the function of the putative regulatory element in the NESP55 DMR, this element would be predicted to have a role in maintenance rather than establishment of imprinting on the maternal *GNAS* allele. The molecular mechanisms governing the role of this putative element remain currently undefined.

Nearly all sporadic PHP-Ib cases show broad epigenetic defects at *GNAS*, including, in each case, a loss of exon A/B imprinting.[65,74] Deletions that involve the entire NESP55 DMR have been ruled out in a number of sporadic PHP-Ib cases through the analysis of polymorphisms in this region (M. Bastepe and H. Jüppner, unpublished data). It is thus possible that these cases represent a genetically distinct form of PHP-Ib caused by mutations disrupting other regulatory elements of *GNAS*. Alternatively, some of these cases might represent a recessive form of PHP-Ib, which might result from homozygous inactivating mutations in a gene encoding a *trans*-acting factor. However, since there is no evidence that imprinted loci other than *GNAS* are also affected in these individuals, this putative trans-acting factor would have to be selectively involved in maintaining or establishing imprinting at the *GNAS* locus. Although a complete paternal-only methylation pattern, i.e., gain of imprinting at NESP55 DMR and loss of imprinting at other DMRs, is frequently observed, some DMRs appear normal in a significant portion of sporadic PHP-Ib cases.[65,74] It is possible, as suggested by Liu et al[74] that the epigenetic changes observed in some sporadic PHP-Ib patients occur in a stochastic manner. However, an analysis of imprinting in several such cases has suggested the existence of a correlation between the different DMRs in which epigenetic alterations are present.[74] While this finding may reflect coregulation of these DMRs, it may also suggest that the different epigenetic alterations observed in PHP-Ib patients are caused by distinct (but perhaps related) genetic lesions. In addition to these questions, an unresolved issue entails penetrance with respect to the epigenetic alterations. Data consistent with incomplete penetrance were reported for a single kindred in whom some, but not all, individuals displayed apparently normal *GNAS* methylation despite inheriting, from their affected mothers, the disease associated chromosome 20q haplotype.[79] PTH-resistance was present only in those that exhibited *GNAS* methylation defects. Available molecular data, however, appear insufficient to rule out whether the *GNAS* methylation pattern in these individuals is truly normal or it is partially altered. The latter would, of course, indicate variable expressivity at the molecular level. Further investigations are required to identify the genetic defects in these different forms of PHP-Ib.

All the genetic defects responsible for PHP-Ib appear to result in the loss of *GNAS* exon A/B imprinting. Since disrupted expression of Gsα is the most likely cause of PTH-resistance observed in PHP-Ib patients, it appears that the exon A/B DMR, which lies immediately upstream of the Gsα promoter, is essential for the proper expression of Gsα at least in the renal proximal tubule. Recent studies of mice with targeted ablation of the exon 1A DMR (mouse homolog of exon A/B) have confirmed the importance of this region in the tissue-specific silencing of Gsα.[80,81] Other aspects of the molecular control and physiological consequences of imprinting at the *Gnas* cluster are discussed in the chapters by Peters and Williamson, Frontera et al and Davies et al.

Conclusion

PHP type-I represents an end-organ resistance syndrome in which signaling of various hormones (primarily PTH) is impaired due to genetic defects that affect activity and/or expression of Gsα. Analysis of patients with different PHP-I subtypes have provided remarkable new insights into the understanding of the complex *GNAS* locus that gives rise to Gsα and several other coding and noncoding transcripts. Although a number of genetic defects responsible for different PHP forms have been identified within or close to the *GNAS* locus, molecular mechanisms underlying the various imprinting defects observed in these patients remain incompletely understood. It will be important to identify patients in whom PHP is caused by novel *GNAS* defects, as careful laboratory investigations of those defects will likely further our knowledge of this complex gene and this unique disorder.

Acknowledgements

The author thanks Dr. Robert C. Gensure (Ochsner Clinic Foundation, Tulane University School of Medicine) for his critical review of the text. The author also wishes to acknowledge the funding support of the National Kidney Foundation and the National Institute of Diabetes and Digestive and Kidney Diseases.

References

1. Yu S, Yu D, Lee E et al. Variable and tissue-specific hormone resistance in heterotrimeric G_s protein a-subunit (G_sa) knockout mice is due to tissue-specific imprinting of the G_sa gene. Proc Natl Acad Sci USA 1998; 95:8715-8720.
2. Chen M, Gavrilova O, Liu J et al. Alternative Gnas gene products have opposite effects on glucose and lipid metabolism. Proc Natl Acad Sci USA 2005; 102(20):7386-7391.
3. Germain-Lee EL, Schwindinger W, Crane JL et al. A Mouse Model of Albright Hereditary Osteodystrophy Generated by Targeted Disruption of Exon 1 of the Gnas Gene. Endocrinology 2005; 146(11):4697-4709.
4. Spiegel AM, Weinstein LS. Inherited diseases involving g proteins and g protein-coupled receptors. Annu Rev Med 2004; 55:27-39.
5. Weinstein LS, Yu S, Warner DR et al. Endocrine Manifestations of Stimulatory G Protein alpha-Subunit Mutations and the Role of Genomic Imprinting. Endocr Rev 2001; 22(5):675-705.
6. Potts JT. Parathyroid hormone: past and present. J Endocrinol 2005; 187(3):311-325.
7. Gensure RC, Gardella TJ, Jüppner H. Parathyroid hormone and parathyroid hormone-related peptide and their receptors. Biochem Biophys Res Commun 2005; 328(3):666-678.
8. Ish-Shalom S, Rao LG, Levine MA et al. Normal parathyroid hormone responsiveness of bone-derived cells from a patient with pseudohypoparathyroidism. J Bone Miner Res 1996; 11:8-14.
9. Murray T, Gomez Rao E, Wong MM et al. Pseudohypoparathyroidism with osteitis fibrosa cystica: direct demonstration of skeletal responsiveness to parathyroid hormone in cells cultured from bone. J Bone Miner Res 1993; 8:83-91.
10. Stone M, Hosking D, Garcia-Himmelstine C et al. The renal response to exogenous parathyroid hormone in treated pseudohypoparathyroidism. Bone 1993; 14:727-735.
11. Breslau NA, Weinstock RS. Regulation of 1,25(OH)$_2$D synthesis in hypoparathyroidism and pseudohypoparathyroidism. Am J Physiol 1988; 255:E730-E736.
12. Drezner MK, Neelon FA, Haussler M et al. 1,25-dihydroxycholecalciferol deficiency: the probable cause of hypocalcemia and metabolic bone disease in pseudohypoparathyroidism. J Clin Endocrinol Metab 1976; 42:621-628.
13. Chase LR, Melson GL, Aurbach GD. Pseudohypoparathyroidism: defective excretion of 3′,5′-AMP in response to parathyroid hormone. J Clin Invest 1969; 48:1832-1844.
14. Albright F, Burnett CH, Smith PH et al. Pseudohypoparathyroidism—an example of "Seabright-Bantam syndrome". Endocrinology 1942; 30:922-932.
15. Farfel Z. Pseudohypohyperparathyroidism-pseudohypoparathyroidism type Ib. J Bone Miner Res 1999; 14:1016.
16. Bastepe M, Jüppner H. GNAS locus and pseudohypoparathyroidism. Horm Res 2005; 63(2):65-74.
17. Levine MA, Downs RW Jr, Moses AM et al. Resistance to multiple hormones in patients with pseudohypoparathyroidism. Association with deficient activity of guanine nucleotide regulatory protein. Am J Med 1983; 74:545-556.
18. Mallet E, Carayon P, Amr S et al. Coupling defect of thyrotropin receptor and adenylate cyclase in a pseudohypoparathyroid patient. J Clin Endocrinol Metab 1982; 54(5):1028-1032.
19. Wolfsdorf JI, Rosenfield RL, Fang VS et al. Partial gonadotrophin-resistance in pseudohypoparathyroidism. Acta Endocrinol (Copenh) 1978; 88(2):321-328.
20. Mantovani G, Maghnie M, Weber G et al. Growth hormone-releasing hormone resistance in pseudohypoparathyroidism type ia: new evidence for imprinting of the Gs alpha gene. J Clin Endocrinol Metab 2003; 88(9):4070-4074.
21. Germain-Lee EL, Groman J, Crane JL et al. Growth hormone deficiency in pseudohypoparathyroidism type 1a: another manifestation of multihormone resistance. J Clin Endocrinol Metab 2003; 88(9):4059-4069.
22. Faull CM, Welbury RR, Paul B et al. Pseudohypoparathyroidism: its phenotypic variability and associated disorders in a large family. Q J Med 1991; 78(287):251-264.
23. Moses AM, Weinstock RS, Levine MA et al. Evidence for normal antidiuretic responses to endogenous and exogenous arginine vasopressin in patients with guanine nucleotide-binding stimulatory protein-deficient pseudohypoparathyroidism. J Clin Endocrinol Metab 1986; 62:221-224.

24. Tsai KS, Chang CC, Wu DJ et al. Deficient erythrocyte membrane Gs alpha activity and resistance to trophic hormones of multiple endocrine organs in two cases of pseudohypoparathyroidism. Taiwan Yi Xue Hui Za Zhi 1989; 88(5):450-455.
25. Weinstein LS, Gejman PV, Friedman E et al. Mutations of the Gs alpha-subunit gene in Albright hereditary osteodystrophy detected by denaturing gradient gel electrophoresis. Proc Natl Acad Sci USA 1990; 87(21):8287-8290.
26. Patten JL, Johns DR, Valle D et al. Mutation in the gene encoding the stimulatory G protein of adenylate cyclase in Albright's hereditary osteodystrophy. New Engl. J Med 1990; 322:1412-1419.
27. Aldred MA, Aftimos S, Hall C et al. Constitutional deletion of chromosome 20q in two patients affected with albright hereditary osteodystrophy. Am J Med Genet 2002; 113(2):167-172.
28. Levine MA, Ahn TG, Klupt SF et al. Genetic deficiency of the alpha subunit of the guanine nucleotide-binding protein Gs as the molecular basis for Albright hereditary osteodystrophy. Proc Natl Acad Sci USA 1988; 85(2):617-621.
29. Patten JL, Levine MA. Immunochemical analysis of the a-subunit of the stimulatory G-protein of adenylyl cyclase in patients with Albright's hereditary osteodystrophy. J Clin Endocrinol Metab 1990; 71:1208-1214.
30. Carter A, Bardin C, Collins R et al. Reduced expression of multiple forms of the a subunit of the stimulatory GTP-binding protein in pseudohypoparathyroidism type Ia. Proc Natl Acad Sci USA 1987; 84:7266-7269.
31. Farfel Z, Brickman AS, Kaslow HR et al. Defect of receptor-cyclase coupling protein in pseudohypoparathyroidism. N Engl J Med 1980; 303:237-242.
32. Levine MA, Downs RW Jr, Singer M et al. Deficient activity of guanine nucleotide regulatory protein in erythrocytes from patients with pseudohypoparathyroidism. Biochem Biophys Res Commun 1980; 94:1319-1324.
33. Linglart A, Carel JC, Garabedian M et al. GNAS1 Lesions in Pseudohypoparathyroidism Ia and Ic: Genotype Phenotype Relationship and Evidence of the Maternal Transmission of the Hormonal Resistance. J Clin Endocrinol Metab 2002; 87(1):189-197.
34. Linglart A, Mahon MJ, Kerachian MA et al. Coding GNAS mutations leading to hormone resistance impair in vitro agonist- and cholera toxin-induced cAMP formation mediated by human XL{alpha}s. Endocrinology 2006; 147(5):2253-62..
35. Farfel Z, Brothers VM, Brickman AS et al. Pseudohypoparathyroidism: inheritance of deficient receptor-cyclase coupling activity. Proc Natl Acad Sci USA 1981;78(5):3098-3102.
36. Albright F, Forbes AP, Henneman PH. Pseudo-pseudohypoparathyroidism. Trans Assoc Am Physicians 1952; 65:337-350.
37. Levine MA, Jap TS, Mauseth RS et al. Activity of the stimulatory guanine nucleotide-binding protein is reduced in erythrocytes from patients with pseudohypoparathyroidism and pseudohypoparathyroidism: Biochemical, endocrine and genetic analysis of Albright's hereditary osteodystrophy in six kindreds. J Clin Endocrinol Metab 1986; 62:497-502.
38. Davies AJ, Hughes HE. Imprinting in Albright's hereditary osteodystrophy. J Med Genet 1993; 30:101-103.
39. Wilson LC, Oude-Luttikhuis MEM, Clayton PT et al. Parental origin of Gsa gene mutations in Albright's hereditary osteodystrophy. J Med Genet 1994; 31:835-839.
40. Kaplan FS, Shore EM. Progressive osseous heteroplasia. J Bone Miner Res 2000; 15(11):2084-2094.
41. Eddy MC, De Beur SM, Yandow SM et al. Deficiency of the alpha-subunit of the stimulatory G protein and severe extraskeletal ossification. J Bone Miner Res 2000; 15(11):2074-2083.
42. Shore EM, Ahn J, Jan de Beur S et al. Paternally inherited inactivating mutations of the GNAS1 gene in progressive osseous heteroplasia. N Engl J Med 2002; 346(2):99-106.
43. Ahmed SF, Barr DG, Bonthron DT. GNAS1 mutations and progressive osseous heteroplasia. N Engl J Med 2002; 346(21):1669-1671.
44. Bastepe M, Weinstein LS, Ogata N et al. Stimulatory G protein directly regulates hypertrophic differentiation of growth plate cartilage in vivo. Proc Natl Acad Sci USA 2004; 101(41):14794-14799.
45. Sakamoto A, Chen M, Kobayashi T et al. Chondrocyte-specific knockout of the G protein G(s)alpha leads to epiphyseal and growth plate abnormalities and ectopic chondrocyte formation. J Bone Miner Res 2005; 20(4):663-671.
46. Mantovani G, Ballare E, Giammona E et al. The Gsalpha Gene: Predominant Maternal Origin of Transcription in Human Thyroid Gland and Gonads. J Clin Endocrinol Metab 2002; 87(10):4736-4740.
47. Germain-Lee EL, Ding CL, Deng Z et al. Paternal imprinting of Galpha(s) in the human thyroid as the basis of TSH resistance in pseudohypoparathyroidism type 1a. Biochem Biophys Res Commun 2002; 296(1):67-72.

48. Liu J, Erlichman B, Weinstein LS. The stimulatory G protein a-subunit Gsa is imprinted in human thyroid glands: implications for thyroid function in pseudohypoparathyroidism types 1A and 1B. J Clin Endocrinol Metabol 2003; 88(9):4336-4341.
49. Hayward B, Barlier A, Korbonits M et al. Imprinting of the G(s)alpha gene GNAS1 in the pathogenesis of acromegaly. J Clin Invest 2001; 107:R31-36.
50. Mantovani G, Bondioni S, Locatelli M et al. Biallelic expression of the Gsalpha gene in human bone and adipose tissue. J Clin Endocrinol Metab 2004; 89(12):6316-6319.
51. Zheng H, Radeva G, McCann JA et al. Gas transcripts are biallelically expressed in the human kidney cortex: implications for pseudohypoparathyroidism type Ib. J Clin Endocrinol Metab 2001; 86(10):4627-4629.
52. Tsang R, Venkataraman P, Ho M et al. The development of pseudohypoparathyroidism. Involvement of progressively increasing serum parathyroid hormone concentrations, increased 1,25-dihydroxyvitamin D concentrations and 'migratory' subcutaneous calcifications. Am J Dis Child 1984; 138:654-658.
53. Barr DG, Stirling HF, Darling JA. Evolution of pseudohypoparathyroidism: an informative family study. Arch Dis Child 1994; 70(4):337-338.
54. Linglart A, Gensure RC, Olney RC et al. A Novel STX16 Deletion in Autosomal Dominant Pseudohypoparathyroidism Type Ib Redefines the Boundaries of a cis-Acting Imprinting Control Element of GNAS. Am J Hum Genet 2005; 76(5):804-814.
55. Jüppner H, Schipani E, Bastepe M et al. The gene responsible for pseudohypoparathyroidism type Ib is paternally imprinted and maps in four unrelated kindreds to chromosome 20q13.3. Proc Natl Acad Sci USA 1998; 95:11798-11803.
56. Bastepe M, Pincus JE, Sugimoto T et al. Positional dissociation between the genetic mutation responsible for pseudohypoparathyroidism type Ib and the associated methylation defect at exon A/B: evidence for a long-range regulatory element within the imprinted GNAS1 locus. Hum Mol Genet 2001; 10:1231-1241.
57. Bastepe M, Lane AH, Jüppner H. Paternal uniparental isodisomy of chromosome 20q (patUPD20q)—and the resulting changes in GNAS1 methylation—as a plausible cause of pseudohypoparathyroidism. Am J Hum Genet 2001; 68:1283-1289.
58. Silve C, Santora A, Breslau N et al. Selective resistance to parathyroid hormone in cultured skin fibroblasts from patients with pseudohypoparathyroidism type Ib. J Clin Endocrinol Metab 1986; 62:640-644.
59. Schipani E, Weinstein LS, Bergwitz C et al. Pseudohypoparathyroidism type Ib is not caused by mutations in the coding exons of the human parathyroid hormone (PTH)/PTH-related peptide receptor gene. J Clin Endocrinol Metab 1995; 80:1611-1621.
60. Fukumoto S, Suzawa M, Takeuchi Y et al. Absence of mutations in parathyroid hormone (PTH)/PTH-related protein receptor complementary deoxyribonucleic acid in patients with pseudohypoparathyroidism type Ib. J Clin Endocrinol Metab 1996; 81:2554-2558.
61. Ding CL, Usdin TB, Labuda M et al. Molecular genetic analysis of pseudohypoparathyroidism type Ib: exclusion of the genes encoding the type 1 and type 2 PTH receptors. J Bone Miner Res 1996; 11(suppl 1):M483.
62. Bettoun JD, Minagawa M, Kwan MY et al. Cloning and characterization of the promoter regions of the human parathyroid hormone (PTH)/PTH-related peptide receptor gene: analysis of deoxyribonucleic acid from normal subjects and patients with pseudohypoparathyroidism type Ib. J Clin Endocrinol Metab 1997; 82:1031-1040.
63. Fukumoto S, Suzawa M, Kikuchi T et al. Cloning and characterization of kidney-specific promoter of human PTH/PTHrP receptor gene: absence of mutation in patients with pseudohypoparathyroidism type Ib. Mol Cell Endocrinol 1998; 141:41-47.
64. Jan De Beur SM, O'Connell JR, Peila R et al. The pseudohypoparathyroidism type lb locus is linked to a region including GNAS1 at 20q13.3. J Bone Miner Res 2003; 18(3):424-433.
65. Liu J, Litman D, Rosenberg M et al. A GNAS1 imprinting defect in pseudohypoparathyroidism type IB. J Clin Invest 2000; 106:1167-1174.
66. Liu J, Yu S, Litman D et al. Identification of a methylation imprint mark within the mouse gnas locus. Mol Cell Biol 2000; 20:5808-5817.
67. Swaroop A, Agarwal N, Gruen JR et al. Differential expression of novel Gs alpha signal transduction protein cDNA species. Nucleic Acids Res 1991; 19(17):4725-4729.
68. Dou D, Joseph R. Cloning of human neuronatin gene and its localization to chromosome-20q 11.2-12: the deduced protein is a novel 'proteolipid'. Brain Res 1996; 723(1-2):8-22.
69. Kikyo N, Williamson CM, John RM et al. Genetic and functional analysis of neuronatin in mice with maternal or paternal duplication of distal Chr 2. Dev Biol 1997; 190(1):66-77.
70. Kagitani F, Kuroiwa Y, Wakana S et al. Peg5/Neuronatin is an imprinted gene located on sub-distal chromosome 2 in the mouse. Nucleic Acids Res 1997; 25(17):3428-3432.

71. Bastepe M, Frohlich LF, Hendy GN et al. Autosomal dominant pseudohypoparathyroidism type Ib is associated with a heterozygous microdeletion that likely disrupts a putative imprinting control element of GNAS. J Clin Invest 2003; 112(8):1255-1263.
72. Laspa E, Bastepe M, Jüppner H et al. Phenotypic and molecular genetic aspects of pseudohypoparathyroidism type ib in a Greek kindred: evidence for enhanced uric acid excretion due to parathyroid hormone resistance. J Clin Endocrinol Metab 2004;89(12):5942-5947.
73. Mahmud FH, Linglart A, Bastepe M et al. Molecular diagnosis of pseudohypoparathyroidism type Ib in a family with presumed paroxysmal dyskinesia. Pediatrics 2005; 115(2):e242-244.
74. Liu J, Nealon JG, Weinstein LS. Distinct patterns of abnormal GNAS imprinting in familial and sporadic pseudohypoparathyroidism type IB. Hum Mol Genet 2005; 14(1):95-102.
75. Simonsen A, Bremnes B, Ronning E et al. Syntaxin-16, a putative Golgi t-SNARE. Eur J Cell Biol 1998; 75(3):223-231.
76. Tang BL, Low DY, Lee SS et al. Molecular cloning and localization of human syntaxin 16, a member of the syntaxin family of SNARE proteins. Biochem Biophys Res Commun 1998; 242(3):673-679.
77. Bastepe M, Frohlich LF, Linglart A et al. Deletion of the NESP55 differentially methylated region causes loss of maternal GNAS imprints and pseudohypoparathyroidism type-Ib. Nat Genet 2005; 37(1):25-37.
78. Coombes C, Arnaud P, Gordon E et al. Epigenetic properties and identification of an imprint mark in the Nesp-Gnasxl domain of the mouse Gnas imprinted locus. Mol Cell Biol 2003; 23(16):5475-5488.
79. Jan de Beur S, Ding C, Germain-Lee E et al. Discordance between genetic and epigenetic defects in pseudohypoparathyroidism type 1b revealed by inconsistent loss of maternal imprinting at GNAS1. Am J Hum Genet 2003; 73(2):314-322.
80. Liu J, Chen M, Deng C et al. Identification of the control region for tissue-specific imprinting of the stimulatory G protein alpha-subunit. Proc Natl Acad Sci USA 2005; 102(15):5513-5518.
81. Williamson CM, Ball ST, Nottingham WT et al. A cis-acting control region is required exclusively for the tissue-specific imprinting of Gnas. Nat Genet 2004; 36(8):894-899.

CHAPTER 4

Imprinted Genes, Postnatal Adaptations and Enduring Effects on Energy Homeostasis

Margalida Frontera, Benjamin Dickins, Antonius Plagge and Gavin Kelsey*

Abstract

The effects of imprinted genes on fetal growth and development have been firmly established. By and large, their roles conform to a conflict over provision of limited maternal resources to offspring, such that paternally expressed imprinted genes in offspring generally promote growth of the fetus, while maternally expressed imprinted genes tend to restrict it. It is comparatively recently that the important effects of imprinted genes in postnatal physiology have begun to be demonstrated, although a similar conflict may apply. In this chapter, we shall review some of the genetic evidence for imprinted effects on obesity, consider the action of selected imprinted genes in the central and peripheral control of energy homeostasis and look in detail at the intriguing effects of imprinting at the *Gnas* locus. Finally, we shall discuss whether these observations fit expectations of the prevailing theory for the existence of imprinting in mammals and go on to consider imprinted genes as targets for developmental programming.

Introduction

The incidence of obesity is increasing rapidly both in industrialized and developing countries, reaching epidemic proportions; it is also a major risk factor for type 2 diabetes mellitus, hypertension and cardiovascular diseases.[1] Obesity is a complex, multifactorial syndrome that is influenced by genetic as well as environmental factors. Despite the obvious supposition that the recent rise in obesity has predominantly environmental and nutritional causes, around 40% of obesity can be attributed to hereditary factors.[2] In as much as imprinted genes have significant effects on postnatal metabolism, it is worth considering a specific role of imprinted genes—or deregulation of imprinted genes—as a contributing genetic factor in obesity. Indeed, obesity occurs in several human imprinting disorders, notably Prader-Willi syndrome (PWS), Angelman syndrome (AS) and Albright's hereditary osteodystrophy (AHO).

Imprinted Gene Syndromes and Obesity

PWS is the most common genetic form of obesity, characterized by life-threatening hyperphagia from childhood[3] together with short stature, low lean body mass, muscular hypotonia, mild mental retardation, behavioral abnormalities and dysmorphic features.[4] Hyperphagia in these patients is accompanied by a massive accumulation of adipose tissue, which shows an unusual fat patterning,[5] suggesting abnormalities in fat mobilization and oxidation or triglyceride synthesis and storage. Impaired adipose metabolism is also indicated by the presence in PWS individuals of increased

*Corresponding Author: Gavin Kelsey—Laboratory of Developmental Genetics and Imprinting, The Babraham Institute, Cambridge, CB22 3AT, U.K. Email: gavin.kelsey@bbsrc.ac.uk

Genomic Imprinting, edited by Jon F. Wilkins. ©2008 Landes Bioscience and Springer Science+Business Media.

lipoprotein lipase activity levels (involved in higher efficiency of triglyceride storage)[6] and the fact that PWS patients have lower fat cell numbers but greater fat cell size compared to control obese individuals.[7,8] PWS is caused by loss of function of a set of paternally expressed imprinted genes in the long arm of chromosome 15 (15q11-q13) (Fig. 1). Approximately 70-75% of PWS cases are associated with 4- to 5- Mb deletions that encompass the imprinted genes and a set of five or nine non-imprinted loci;[9,10] 20-25% with maternal uniparental disomy (UPD) of chromosome 15; 2-5% with imprinting mutations and 1% with translocations.[11] As point mutations in a single gene have not been found in patients, it has been proposed that PWS is a contiguous gene syndrome, resulting from the loss of expression of more than one paternally expressed gene in the region. The PWS imprinted candidate genes include three intronless genes (*NDN*, *MAGEL2* and *MKRN3*), a complex polycistronic locus (*SNURF-SNRPN*) and various snoRNAs.[9] The pathophysiological cause of the hyperphagia observed in PWS patients is unclear, though it is felt to be hypothalamic in origin;[3] candidate genes for PWS are indeed expressed in specific areas of the hypothalamus involved in the regulation of energy balance.[12,13] PWS individuals show abnormally elevated levels of plasma ghrelin,[14-17] an orexigenic gut hormone that is usually found at low levels in obesity. Ghrelin stimulates feeding through the growth hormone-secretatogue receptor (GHS-R), at least partially through activation of neuropeptide Y (NPY) and agouti-related peptide (AGRP) in hypothalamic arcuate nucleus neurons.[18] Chronic peripheral or central administration of ghrelin to rodents causes obesity[18,19] and ghrelin acutely stimulates appetite when infused in humans.[20] Thus, it has been hypothesized that the hyperghrelinemia found in PWS patients could contribute to their hyperphagia and obesity,[14,15] however, forced reduction of ghrelin plasma levels in PWS-affected individuals by administration of pharmacological agents does not produce a concomitant reduction in food intake.[21] Because the *ghrelin* gene is not linked to the PWS region, *hyperghrelinemia* in these subjects must be secondary to deregulation of one or more transcripts in the PWS locus. Interestingly, there is some evidence that ghrelin secretion is regulated in part by the parasympathetic system,[22] and a possible connection between altered ghrelin levels in PWS patients and a hypothalamic defect affecting parasympathetic activity has been suggested.[23]

A second imprinting disorder linked to obesity is AS; it maps to the same genetic region as PWS in chromosome 15q11-q13 and can be regarded genetically as the reciprocal disorder. AS is characterized by severe mental retardation, ataxia, seizures with EEG abnormalities, subtle dysmorphic facial features and an apparently happy, sociable disposition. Like PWS, there are multiple genetic causes of AS, but the key finding is the failure to inherit a normal maternal copy of the gene encoding ubiquitin protein ligase E3A (*UBE3A*), which participates in the ubiquitination of proteins, a process that marks proteins destined for degradation.[24,25] *UBE3A* shows tissue-specific imprinting, being expressed predominantly from the maternal allele in the brain (Purkinje cells, hippocampal neurons and mitral cells of the olfactory bulb). Obesity is a common feature in adult AS patients with *UBE3A* mutations or UPD/imprinting defects,[26] and adult-onset obesity has

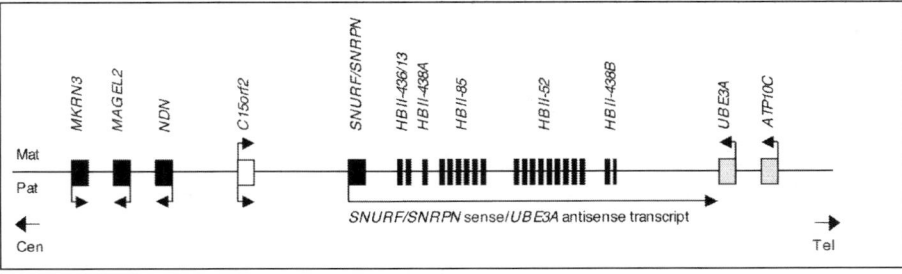

Figure 1. Simplified schematic of the PWS human chromosomal region in 15q11-q13 (not to scale). Paternal expressed genes are drawn in black; maternal expressed genes are in grey; biallelic expressed genes are in open squares. The orientation of transcription for each gene is represented by an arrow. Vertical black bars indicate snoRNA transcripts.

also been described in some AS mouse models.[27,28] Although this suggests that obesity could be due to loss of imprinted *UBE3A* expression, *Ube3a*-deficient mice do not show increased body weight.[29,30] On the other hand, mice with paternal UPD for the AS orthologous region,[28] or a maternally derived deletion immediately proximal to *Ube3a*,[27,31,32] do present with adult-onset obesity. Genetic studies in the mouse have mapped a novel gene very close to *Ube3a*,[31] *Atp10c*, encoding a putative phospholipid translocase that is preferentially expressed from the maternal allele in mouse and human brain.[33,34] The human ortholog, *ATP10C*, maps to the AS critical region in chromosome 15q12.[33-35] Loss of maternal expression of *Atp10c* causes obesity in mice,[36] together with insulin resistance in association with glucose intolerance,[35,37] suggesting that lack of expression of maternal *ATP10C* and/or *UBE3A* could be related to the obesity associated with a certain subset of AS patients.

Genetic Evidence for Parent-of-Origin Effects on Obesity

Evidence for a role of imprinted genes in the regulation of body weight is found in population studies of obesity-related traits and parent-of-origin effects in humans. Lindsay et al.[38] reported a linkage between body mass index (BMI) and parent-of-origin effects in Pima Indians to regions of chromosome 5 and chromosome 10. Another study examining the linkage of BMI to parent-of-origin effects in children, adolescents and young adults[39] identified effects in the youngest sample for chromosomes 3, 4, 10 and 12 and suggested the existence of a 'maternally imprinted' locus in 10p12 that might influence human obesity. More recently, Dong et al.[40] showed in a genome-wide parent-of-origin linkage analysis the existence of three regions in the human genome (10p12, 12q24 and 13q32) that appear to influence obesity when transmitted exclusively from a specific parent. It is suggestive to see a linkage recapitulated in two studies, although none of these regions is known to harbor imprinted genes.[40] Similarly, QTL mapping has suggested the existence of a maternally expressed gene affecting body mass located in mouse chromosome 8,[41] although there is no evidence yet of imprinting of candidate genes in this region. A role for imprinting in body composition in pigs has also been proposed.[42]

Imprinted Gene Action in the Hypothalamus

Imprinted genes could influence the potential for obesity by direct effects in adipose tissues, by action in the central nervous system (CNS), particularly the hypothalamus, or indirectly through other influences on metabolism (Fig. 2). The hypothalamus plays a key role in the integration of physiological processes essential for survival and reproduction; amongst its functions are the regulation of blood pressure, body temperature regulation, energy balance and the expression of sexual and maternal behaviors.[43] An intriguing study that examined the distribution of parthenogenetic (Pg) and androgenetic (Ag) cells in the brain of chimeric mice revealed that imprinted genes may play a role in the correct development of the hypothalamus and other parts of the CNS and that the two parental genomes have different influences (see also Davies et al and Goos and Ragsdale, this volume). Pg cells (disomic for the maternal and nullisomic for the paternal genomes, respectively) in chimeric brains are prevalent in telencephalic structures, including the cortex, striatum and hippocampus and are largely excluded from diencephalic structures, especially the hypothalamus, while Ag cells (disomic for the paternal and nullisomic for the maternal genomes) are found mainly in the hypothalamus, septum, preoptic area and bed nuclei of the stria terminalis, but not in the cortex.[44] The distribution of Ag cells in the hypothalamus is of special relevance in relation to effects of paternal expressed genes in energy metabolism, as this brain region is concerned with neuroendocrine function and feeding regulation. It is also noteworthy that the stage during which these complementary distributions of Pg and Ag cells takes place coincides with the proliferation and differentiation of hypothalamic neural cells, suggesting that paternal expressed genes might be required for proliferation, differentiation or survival of hypothalamic cells.[43] *Peg1/Mest* (paternal expressed gene 1/mesoderm-specific transcript) and *Peg3* (paternal expressed gene 3), two imprinted genes that show effects on body weight regulation, are strongly expressed in those regions where Ag cells accumulate, namely the hypothalamus, preoptic area and septum.[45,46]

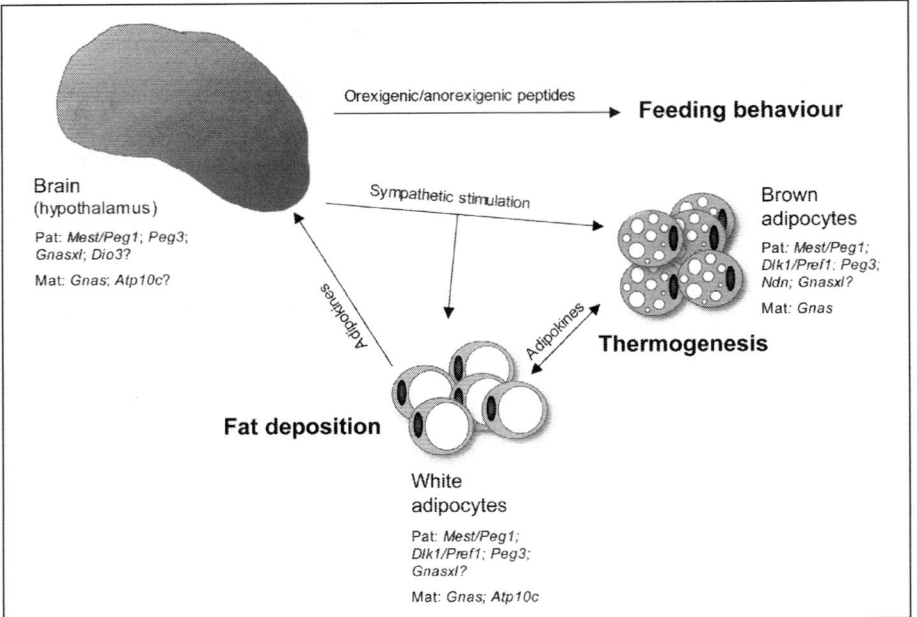

Figure 2. Schematic representation of the brain-adipose tissue axis in the control of energy homeostasis. The brain and especially the hypothalamus, receive and integrate information about the status of the lipid reserves through adipokines released by adipose tissue. Output from the brain will produce a response at different levels; modifying the food intake through the action of orexigenic/anorexigenic peptides; modifying energy expenditure by regulating thermogenesis in BAT; or altering the lipogenic/lipolytic activities in WAT through regulation of sympathetic stimulation. Several imprinted genes (Mat. maternal expressed genes; Pat. paternal expressed genes) appear to act in the regulation of energy homeostasis, both in the brain and adipose tissues.

Peg3 is located in mouse proximal chromosome 7, in the region of conserved synteny with human chromosome 19q13.4.[47] It encodes a C_2H_2-type zinc finger protein that is involved in the regulation of p53-mediated apoptosis.[48,49] It is expressed mainly in placenta, gonads, hypothalamus and adult skeletal muscle, with significant expression in white adipose tissue (WAT) as well and weak expression in brown adipose tissue (BAT).[45,50] Lack of *Peg3* results in developmental delay of the fetus and pups are growth-retarded at birth and have impaired suckling.[51,52] At adult stages, *Peg3* mutants have increased adiposity, despite having a lower body weight and being hypophagic.[50] This effect seems to be a consequence of lower energy expenditure, shown by lower core body temperature and metabolic rate, probably due to alterations in the activation of the sympathetic nervous system (SNS) in the hypothalamus.[50]

Imprinted Gene Action in Adipose Tissues

Although it has been traditionally considered a passive energy reservoir, adipose tissue is a complex and highly active metabolic and endocrine organ that plays a key role in the regulation of energy metabolism. Adipose tissue not only responds to afferent signals from the CNS and other hormone systems, but adipocytes secrete bioactive peptides, termed adipokines, that act locally and distally through autocrine, paracrine and endocrine effects and that are involved in several vital functions, such as appetite and energy balance, lipid metabolism and insulin sensitivity. Furthermore BAT, a special type of adipose tissue, has the unique feature of accumulating triglycerides in order to use them as substrates for the dissipation of energy in the form of heat through a specialized protein, the uncoupling protein 1 (UCP1) (for review, see ref. 53). BAT is the effector organ for nonshivering thermogenesis, an adaptive response to cold temperature that plays an important role in the thermoregulation of newborn mammals, hibernation and protection against diet-induced obesity. A number of imprinted genes may have direct actions in adipocytes.

Adipocyte proliferation and differentiation are strongly related to the development and presence of obesity. A role for three paternal expressed genes, *Pref1/Dlk1* (Preadipocyte factor 1/Delta, Drosophila, Homolog-like 1), *Peg1/Mest* and *Ndn* (Necdin) in adipocyte metabolism has been suggested recently by a number of studies. *Pref1/Dlk1* is located in mouse chromosomes 12, the human ortholog in chromosome 14,[54-56] and codes for a cell-surface transmembrane protein containing epidermal growth factor-like repeats involved in regulating development and differentiation of adipose, mesenchyme, neuroendocrine and hepatopoietic tissues.[57,58] *Pref1/Dlk1* is widely expressed in embryonic tissues, however, it is down-regulated postnatally and in adults it is only found in preadipocytes, pancreatic β-cells, thymocytes and cells in the adrenal gland.[58-61] Paternal transmission of a *Pref1/Dlk1* null allele in mice affects embryonic development and pups at birth are growth-retarded, while the adults become obese due to fat accretion.[62] The opposite phenotype is found in transgenic mice over-expressing *Pref/Dlk1* specifically in adipose tissue: these animals show lower body weight due to decreased fat accumulation, together with abnormal glucose homeostasis and insulin sensitivity.[136] In vitro studies have shown that forced expression of *Pref1/Dlk1* inhibits differentiation of 3T3-L1 adipocyte,[63,64] while its down-regulation promotes adipogenesis.[58,63] It has been proposed that *Pref1/Dlk1* functions in the maintenance of the cells in the preadipocyte state.

The second paternally expressed gene that appears to regulate the adipogenic process is *Peg1/Mest*, located in mouse chromosome 6, the human ortholog in chromosome 7.[65,66] *Peg1/Mest* expression is very low in adult tissues;[67] however its mRNA levels are greatly increased in WAT from mice with high-fat diet induced and genetically induced obesity, probably as a consequence of demethylation of the *Peg1/Mest* promoter in the maternal allele and/or the enhancement or derepression of the *Peg1/Mest* promoter via a methylation-independent mechanism.[68] Ectopic expression of *Peg1/Mest* increases the expression of adipocyte markers, both in vivo and in vitro and transgenic mice over-expressing *Peg1/Mest* are obese and have enlarged adipose cells.[68] Further, suggestive evidence for the involvement of *Peg1/Mest* in body weight regulation comes from the observation that in *Mus* interspecies hybrids there is biallelic expression of *Peg1/Mest* due to loss of imprinting and that this correlates with increases in organs and body weight.[69] Koza et al have very recently reported a striking elevation of *Peg1/Mest* levels in adipose tissue of inbred mice susceptible to diet-induced obesity, highlighting the importance of this gene in the regulation of fat mass expansion.[70]

NDN is a transcriptional regulator of the melanoma-associated antigen (MAGE) protein family coded in the PWS region (Fig. 1). It has been shown to be a growth suppressor controlling neuronal differentiation and survival,[71] but it is also expressed in adipocytes,[72,73] and differentially expressed in white and brown preadipocytes.[73] Recently, Tseng et al demonstrated a role for murine *Ndn* in inhibiting brown adipocyte differentiation, probably via interaction with the E2F family of transcription factors,[74] but mice with a targeted deletion of the *Ndn* gene have not been reported to be obese.[72,75]

Other imprinted genes implicated in the regulation of energy homeostasis from genetic manipulation studies include the paternal expressed genes *Igf2* (insulin-like growth factor 2), *Rasgrf1* (Ras-guanine nucleotide-releasing factor 1) and *Dio3* (type 3 deiodinase); and the maternal expressed gene *Meg1/Grb10* (maternal expressed gene 1/growth factor receptor-bound protein 10) (Table 1). While loss of paternal *Igf2* expression results in embryonic growth deficiency,[76] adult mice lacking *Igf2* expression in the brain owing to deletion of an intergenic enhancer show increased adiposity while being hypophagic,[77] suggesting that *Igf2* could affect postnatal adipocyte metabolism and/or energy expenditure.

Table 1. Role of murine imprinted genes in the control of energy homeostasis: evidence from genetic manipulation studies

Gene	Chromosome	Expressed Allele	Genetic Manipulation	Phenotype	References
Gnasxl	2	Paternal	Knockout	Lean, hypermetabolic, increased insulin sensitivity	97, 108
Gnas	2	Biallelic/Maternal	Knockout	Obese, hypometabolic, insulin resistance	80, 82, 106, 109, 113
Peg1/Mest	6	Paternal	Overexpression	Obese	68
Peg3/Pw1	7	Paternal	Knockout	Obese, hypometabolic	50
Igf2	7	Paternal	Knockout	Obese	77
Atp10c	7	Maternal	Knockout	Obese, insulin resistance, glucose intolerance	37, 132
Rasgrf1	9	Paternal	Knockout	Lean, hypoinsulinemic, impaired glucose tolerance	133
Grb10	11	Maternal	Overexpression	Lean, hyperinsulinemic, insulin resistance	134
Dio3	12	Paternal	Knockout	Lean	135
Dlk1/Pref1	12	Paternal	Overexpression	Lean, hyperinsulinemic, impaired glucose tolerance, insulin resistance	136
			Knockout	Obese	62

The *Gnas* Locus

An imprinted locus for which a role in metabolism and adiposity is particularly well established is *Gnas*. The locus is remarkable for the presence of oppositely imprinted transcripts that encode distinct proteins (Fig. 3) and for the distinct and broadly opposite physiological effects they have. It is likely, also, that this locus influences metabolism at several levels, with effects on food intake, glucose homeostasis, metabolic rate and actions in both the CNS and peripheral tissues. We shall concentrate on the function of the imprinted *Gnas* locus transcripts; Peters and Williamson provide a description on how imprinting of the locus is regulated in a previous chapter.

The authentic *Gnas* transcript encodes the protein Gsα, which together with β- and γ-subunits, forms the trimeric Gs-protein that mediates signal transduction from activated neurotransmitter- and hormone-receptors to adenylate cyclase to stimulate cAMP formation.[78,79] Gsα is widely expressed, but expression is imprinted in a tissue-specific way, such that Gsα is produced only, or predominantly, from the maternal allele in certain tissues and cell types. Among these in mice are BAT and WAT, proximal renal tubules and, possibly, the reproductive system,[80-83] although *Gnas* imprinting in adipose tissue has been disputed.[84] In humans, predominant maternal origin of Gsα expression has been shown for ovary, anterior pituitary (somatotroph cells) and thyroid, but not visceral adipose tissue.[85-90]

The overlapping *Gnasxl* transcript is, in contrast to *Gnas*, exclusively paternally expressed. It encodes an unusual 'extra-large' variant of Gsα, termed XLαs.[91] The XLαs protein possesses a distinct amino-terminus in which the first 46 amino acids of Gsα (encoded by exon 1) are replaced by a >380 'XL' domain encoded by the upstream *Gnasxl* exon that is spliced in frame onto *Gnas* exon 2. The XL domain is not well conserved, although its carboxy-terminal end retains a functional amino acid sequence for interaction with G-protein βγ-subunits.[92] Like Gsα, it also contains

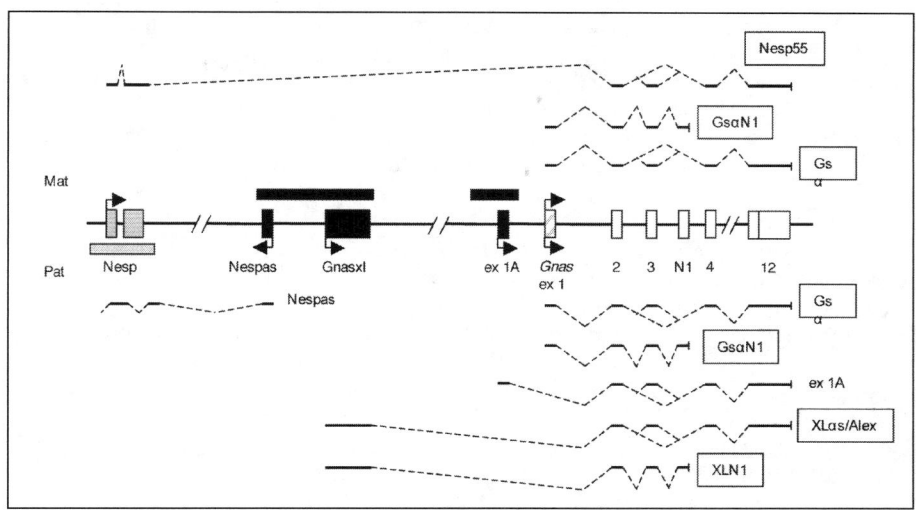

Figure 3. Scheme of the imprinted *Gnas* locus of the mouse. Maternal (Mat) and paternal (Pat) allele-specific features are indicated in black and grey, respectively. Initiation of maternal- and paternal-specific transcripts is shown by arrows and their splice patterns are given above and below. For simplicity, *Gnas* exons 5-11 are omitted. Tissue-specific maternal expression of the *Gnas* promoter (coding for Gsα) is represented by the striped box (ex 1). The names of protein coding transcripts are boxed. GsαN1 is a neural-specific truncated form of Gsα. *Gnasxl* encodes XLαs, an amino-terminal variant of Gsα, a neural-specific truncated protein XLN1 and the unrelated protein Alex. The Nesp55 protein is coded in a single upstream Nesp exon. *Nespas* and *exon 1A* (ex 1A) transcripts produce noncoding RNAs. Regions of imprinted methylation (DMRs) are indicated by bars above or below exons (adjusted from ref. 97, with permission from Nature Publishing Group).

several cysteine residues that mediate lipid-anchorage to the cytosolic side of the plasma membrane via palmitoylation.[93] XLαs can mediate signalling from activated receptors (e.g., PTH-R1, TSH-R, CRF-R1, β_2-AR) to stimulate adenylate cyclase similar to Gsα when re-introduced into fibroblasts that are genetically deficient for both proteins.[94,95] Alternative splicing onto exon N1 (located between *Gnas* exons 3 and 4) in brain, pituitary gland and adrenal medulla produces a truncated version of XLαs, called XLN1, such that the amounts of XLN1 mRNA and protein exceed those of full length XLαs in these tissues.[96,97] The function of the XLN1 protein is unknown; it retains the Gsα-like domains for membrane anchorage and βγ-subunit interaction, but lacks all of the remaining sequences for GTP binding and interaction with receptors and adenylate cyclase. A unique property of the *Gnasxl* exon is the presence of a second conserved open reading frame that begins downstream of the XLαs start codon and is shifted by +1 nucleotide relative to the XLαs frame.[92] Alex, the protein produced from this overlapping reading frame, terminates at the end of the *Gnasxl* exon and is unrelated in sequence to any G-protein. Occurrence of a second, frame-shifted translation initiation from the same mRNA is unprecedented in mammals, but this relationship is conserved, despite rapid evolution of XLαs and Alex coding sequences,[98] and Alex protein has been detected in rat PC12 cells and in human platelets.[92,99] More surprising still is the finding that Alex and XLαs bind to each other in vitro and can be co-immunoprecipitated from human platelets. That the interaction may be functionally important is suggested by the fact that it is impaired in patients carrying a sequence polymorphism that affects both proteins and results in reduced platelet aggregation.[99,100] What effect Alex and its proposed interaction with XLαs have on metabolism is not known.

A further protein, the chromogranin-related protein Nesp55, is produced from the *Gnas* locus and is translated from a separate transcript that begins at a distant upstream promoter, but shares downstream exons with other transcripts of the locus.[101] Nesp55 is only expressed from maternal allele and very little is known about its function at the molecular level. As knock-out work in the mice reveals no function in common with Gsα and XLαs, nor obvious effects on metabolism,[102] we shall not consider Nesp55 further here (but see the chapter by Davies et al in this volume).

The Role of Gsα in Energy Homeostasis

From a number of recent studies using genetically engineered mice, we now know that the proteins derived from the *Gnas* and *Gnasxl* transcripts have crucial roles in some of the novel physiological functions that mammalian offspring need to establish after birth and have sustained effects on metabolism in adults. Furthermore, their opposite patterns of imprinted expression are mirrored by largely opposite phenotypes in the respective mouse mutants (Table 2), pointing to antagonistic roles of Gsα and XLαs/XLN1/Alex in these physiological pathways. We shall concern ourselves mainly with the function of this locus in metabolism; further information on the complexity of functions at this locus can be found in references 103, 104.

Although mutations of the *Gnas* locus in the mouse have been in existence for some years, they potentially affect several transcripts,[80,105] so that it was first through specific deletion of *Gnas* exon 1 that the role of Gsα could be determined unequivocally.[82,84] While complete loss of Gsα in homozygous knock-out mice results in early embryonic lethality, heterozygous mutants show haploinsufficiencies differing in several respects dependent on parental origin of the mutant allele, clear evidence for the consequences of imprinted *Gnas* expression.

If the mutation is inherited on the maternal allele (exon 1 m–/p+), mice are born with severe subcutaneous edema[82] (as previously observed in *Gnas* exon 2 m–/p+ and *Oed* mutants)[105,106] that may help account for a high rate of postnatal losses (50-60%) and which declines during the first 2-3 days. Adult mice with lack of maternal allele-specific Gsα expression develop a metabolic phenotype and deregulated energy homeostasis.[82,84] Their increased body weight (+20%) is due to increased adipose tissue mass and lipid accumulation. However, this is not caused by increased food intake, but is correlated with a reduced metabolic rate and a tendency towards lower activity levels. Other correlates of obesity are also found in these mice, i.e., hyperglycemia, hyperinsulinemia with insulin resistance and glucose intolerance, hyperlipidemia and hyperleptinemia.[82] The cause for the increased adiposity of *Gnas* m–/p+ mice is not entirely clear, but it has been hypothesized that reduced activity of the SNS is involved, resulting in reduced energy expenditure. This inference is based on indirect indicators of SNS activity, e.g., lower levels of norepinephrine in urine and reduced levels of the uncoupling protein UCP1 in BAT, which is mainly regulated by the SNS via β-adrenergic receptors.[106] Alternatively or additionally, loss of Gsα in adipocytes (owing to its imprinting) might lead to a reduced response to β-adrenergic stimulation and a low level of lipid mobilization.

Heterozygous mice with a lack of paternal allele-specific Gsα expression (exon 1 m+/p–) show normal postnatal development on an outbred background,[82] or 31-40% mortality in the inbred 129SvEv strain[84] (viability of many *Gnas* locus mutants is reduced on inbred backgrounds). Adult exon 1 m+/p– mice display some phenotypic features reminiscent of exon 1 m–/p+ mice, but these are much less severe. Thus, they also develop increased amounts of adipose tissue and mild hyperinsulinemia, insulin resistance and glucose intolerance,[82] but there is no evidence of abnormal metabolic rate, locomotor activity or SNS function, pointing to a different origin of these physiological phenotypes in paternal versus maternal mutants. It may be that reduced adipose expression of Gsα in exon 1 m–/p+ and m+/p– mice (to differing degrees because of partial imprinting) promotes lipid storage through partial resistance to the lipolytic actions of norepinephrine, but that in exon 1 m–/p+ mice there are additive effects of reduced metabolic rate/SNS activity that cause more extreme obesity. A protein with a role as central to metabolism as Gsα clearly has pleiotropic effects, so that tissue-specific knock-outs become essential for identifying tissue involvement in a phenotypic endpoint as multifactorial as obesity. For example, homozygous loss of Gsα specifically

Table 2. *Metabolism, feeding and energy homeostasis related characteristics of mouse models of the Gnas locus*

		Lack of Paternal Function			
	MatDp(dist2)	Gnas Exon 1[m+/p−]	Gnas Exon 2[m+/p−]	Sml	Gnasxl[m+/p−]
	Gsα expression increased in imprinted tissues; XLαs/XLN1/Alex absent	Gsα expression reduced 50% in non-imprinted tissues	Gsα expression reduced 50% in non-imprinted tissues; XLαs/XLN1/Alex absent	Point mutation in Gnas exon 6; Gsα function reduced 50% in non-imprinted tissues; XLαs function lost	XLαs/XLN1/Alex absent
	Inability to suckle 100% lethality within 24 h	Normal survival and normal preweaning phenotype[82]; 31–40% lethality[84]	Reduced suckling 77% preweaning lethality	Substantial preweaning loss	Reduced suckling 80–100% preweaning lethality
	Decreased adiposity (BAT)*		Decreased adiposity (BAT and WAT)	Decreased adiposity (BAT)*	Decreased adiposity (BAT and WAT); increased cAMP in BAT Hypoglycemia; hypoinsulinemia; hypoglucagonemia
			Postnatal growth retardation	Postnatal growth retardation	Postnatal growth retardation
		Adults: obese; hyperinsulinemia; glucose intolerant; insulin resistant; reduced triglyceride clearance	Adults: lean; reduced body weight; reduced lipid accumulation in BAT and WAT; hypoinsulinemia; hypermetabolic; increased locomotor activity; increased glucose tolerance; increased insulin sensitivity; increased glucose uptake into muscle,	Adults: lean*	Adults: lean; reduced body weight; reduced lipid accumulation in BAT and WAT; greater mitochondrial content in BAT; hypoglycemia; hypoinsulinemia; hypolipidemia; hypermetabolic; increased glucose tolerance; increased

continued on next page

Table 2. Continued

	Lack of Paternal Function		
MatDp(dist2)	Gnas Exon 1m+/p−	Gnas Exon 2m+/p−	Sml	Gnasxlm+/p−
		WAT and BAT; increased triglyceride clearance		insulin sensitivity, signalling and turn-over; increased glucose uptake into muscle, WAT and BAT; increased triglyceride clearance; hyperphagia; increased lipolysis; increased sympathetic tone; increased expression of genes involved in lipid oxidation, thermogenesis, mitochondrial function and adipogenesis in adipose tissue; reduced expression of genes involved in lipogenesis in liver.
Refs. 137, 138	82, 84	80, 106, 109, 113	105	97, 108

	Lack of Maternal Function	
PatDp(dist2)	Gnas Exon 1m−/p+	Gnas Exon 2m−/p+	Oed
Gsα expression reduced in imprinted tissues; XLαs/XLN1/Alex expression increased by a factor of 2	Gsα expressed reduced 50% in imprinted tissues and >50% in imprinted tissues	Gsα expressed reduced 50% in non-imprinted tissues and >50% in imprinted tissues	Point mutation in Gnas exon 6; Gsα function reduced 50% in non-imprinted tissues and >50% in imprinted tissues

continued on next page

Table 2. Continued

	Lack of Maternal Function		
PatDp(dist2)	**Gnas Exon 1m−/p+**	**Gnas Exon 2m−/p+**	**Oed**
100% perinatal lethality	49%[82] or 66%[84] preweaning lethality	80% preweaning lethality	Substantial preweaning lethality
Increased adiposity (BAT)*	Adults: obese; hyperglycemia; hyperinsulinemia; hyperlipidemia; hypometabolic; decreased locomoter activity; glucose intolerant; insulin resistant	Increased adiposity (BAT and WAT) Adults: obese; hypometabolic; decreased locomoter activity; increased insulin sensitivity;	Increased adiposity (BAT)
Refs. 137, 138	82, 84	80, 106, 109, 113	105

Homozygous Loss of Function
Gnas Exon 1−/− (Liver Specific)
Gsα absent in liver Adults: resistance to glucagon signalling and glycogen breakdown; increased liver glycogen content; pancreatic islet hyperplasia; very high levels of serum glucagon and

continued on next page

Table 2. Continued

	Homozygous Loss of Function
Gnas Exon 1⁻/⁻ (Liver Specific)	GLP-1; hypoglycemia; hypoinsulinemia; reduced expression of genes involved in gluconeogenesis; increased glucose tolerance and insulin sensitivity; increased uptake of glucose into liver and muscle; reduced adipose tissue mass; increased lipolysis (serum free fatty acid levels); increased liver lipogenesis (triglyceride secretion) also under fasting conditions
	ref. 107

* Jo Peters, unpublished observations

in liver (where expression is biallelic) causes profound hepatic glucagon resistance (whose receptor is Gs-coupled), with knock-on effects on glucose homeostasis (improved glucose tolerance and uptake in liver and muscle) and deregulated lipid metabolism, with reduced fat mass and increased serum free fatty acid levels but increased hepatic lipogenesis.[107] An adipocyte-specific ablation of Gsα is eagerly awaited, as final proof of the degree of imprinting of Gsα in BAT and WAT and the extent of its physiological significance.

Apart from effects on energy homeostasis, exon 1 m–/p+ mice uniquely display other physiological impairments. Cells with imprinted *Gnas* expression lose 50-100% of Gsα when a mutation is inherited maternally, depending on the degree of paternal allele silencing and reduced receptor coupling can lead to resistance to certain hormones. Systems that are affected in this way include the parathyroid hormone regulation of blood Ca^{2+} and phosphate levels (see the chapter by Bastepe), thyrotropin regulation of the thyroid gland as well as fertility and maternal reproductive success;[84] also reviewed in refs. 103, 104. Comparison of maternally and paternally inherited haploinsufficiencies of Gsα thus help to reveal the degree of *Gnas* imprinting in physiological systems, such that the greater the phenotypic discrepancy between exon 1 m–/p+ and $^{m+/p-}$ mice, the greater the extent of suppression of the paternal *Gnas* allele.

XLαs in Postnatal Adaptations and Metabolism

Mice deficient in the paternal allele-derived *Gnasxl* transcript reveal many effects opposite to those of maternal *Gnas* mutants, especially with regard to metabolism and energy homeostasis. A specific disruption of the *Gnasxl* exon results in loss of the three proteins it encodes (XLαs, XLN1 and Alex) without affecting Gsα expression.[97] (Therefore, this knock-out does not resolve the separate functions of the three proteins.) *Gnasxl* $^{m+/p-}$ pups exhibit a failure-to-thrive phenotype: inertness, reduced suckling activity and growth retardation. Few survive beyond weaning (~10% on crosses to outbred CD1, somewhat more if intra-litter competition is reduced) and these survivors show a 50-60% lower body weight. The deficit in energy resources becomes obvious soon after birth. Mutant pups are hypoglycemic and concomitantly hypoinsulinemic at all postnatal stages. Adipose tissue is significantly reduced in proportion and depleted of lipid stores. BAT was found to contain elevated levels of cAMP in *Gnasxl* $^{m+/p-}$ pups,[97] compatible with hyperactivity of the β-adrenergic signaling pathway, which would also result in increased UCP1 protein levels.[106] It is currently unresolved whether increased BAT cAMP content is entirely a consequence of increased stimulation by the SNS or whether effects intrinsic to the adipocyte also contribute, since *Gnasxl* is normally expressed in this tissue at early postnatal stages.[106,108] There is no information on the BAT cAMP content of *Gnas* exon 1 m–/p+ pups for comparison.

The early postnatal expression pattern of the *Gnasxl* transcript is highly specific, in contrast to the widespread expression of *Gnas* and provides suggestive indications of the tissues and cell types involved in the *Gnasxl*-deficiency phenotype. Its distribution in the brain is restricted mainly to defined brainstem regions,[96,97] for example, loss of expression from the three motornuclei that innervate the tongue and orofacial muscles (facial, motor-trigeminal and hypoglossal nuclei), could contribute to suckling deficiency. The locus coeruleus, which is the main noradrenergic centre of the brain and regulates states of alertness and influences the processing of many kinds of sensory information, including a specific role in neonatal olfactory learning, expresses high levels of *Gnasxl*. *Gnasxl* is also found in scattered cells of the hypothalamus and medulla oblongata, which might constitute part of the SNS. Other sites of expression potentially relevant to metabolism include parts of the endocrine system such as the anterior and intermediate part of the hypophysis and the adrenal medulla, as well as some peripheral tissues, e.g., BAT and WAT, pancreas.[96,97] The expression of *Gnasxl* is dynamic and changes towards adulthood. *Gnasxl* ceases to be expressed around weaning age in adipose tissue[108] and it is likely that other tissues undergo changes in expression: in contrast to neonates, adult mutants have no difficulties in feeding and show increased food intake[108].

The contrasting metabolic phenotypes of lack of paternally expressed XLαs/XLN1/Alex versus maternally expressed Gsα become fully apparent after weaning. *Gnasxl* mutants remain underweight throughout life with reduced amounts of adipose tissue and lipid resources, despite evidence of hyperphagia and have low levels of circulating leptin. They are also hypermetabolic and have increased energy expenditure,[108] low blood glucose and insulin levels combined with increased glucose tolerance and insulin sensitivity, which leads to enhanced glucose uptake into muscle and adipose tissue. Elevated levels of norepinephrine in urine hint at an increased sympathetic tone. A number of gene expression changes have been identified in adipose tissue, mostly comprising up-regulation of genes that promote thermogenesis, adipogenesis, mitochondrial biogenesis and lipid oxidation. Since *Gnasxl* is not expressed in adult adipose tissues, these phenotypic features might be a consequence of a continuous hyperstimulation by the SNS. However, a number of questions remain unanswered, e.g., how could *Gnasxl* derived proteins down-regulate SNS outflow and whether SNS regulation of other peripheral tissues is changed as well. It also remains open whether lack of *Gnasxl* in neonatal adipose tissue has enduring effects on development and differentiation of BAT and WAT towards adulthood.

All in all, it is quite remarkable that the effects of *Gnasxl* deficiency on adult metabolism are broadly opposite to those of loss of maternally expressed *Gnas*. *Gnasxl* deficiency also appears to be dominant over *Gnas* haploinsufficiency, because the *Gnasxl*-null phenotype is essentially identical to that of mice with a paternally inherited mutation in *Gnas* exon 2, which lack paternal expression of *Gnas* as well as *Gnasxl*.[80,106,109] The challenge now is to elucidate at what level the physiological antagonism is established, whether Gsα and XLαs/XLN1/Alex act in separate, opposite physiological regulatory pathways in vivo and whether and how XLN1 and Alex contribute to the phenotype observed in *Gnasxl* $^{m+/p-}$ mice.

Mutations of the *GNAS* Locus in Human Neonatal Physiology and Adult Energy Homeostasis

With the high degree of conservation of Gsα and the very similar organization the *GNAS* locus and its imprinting in humans as compared with mice, one might expect similar consequences of mutations and this is true in part. Inactivating mutations of human *GNAS* cause AHO. Consistent with tissue-specific imprinting of human *GNAS*, maternal inheritance of mutations additionally causes a hormone-resistance syndrome called pseudohypoparathyroidism (PHP)[78] (see also the chapter by Bastepe). AHO has quite variable presentation, but one common feature is obesity from childhood onwards, although this particular aspect has not been a main focus of research. Most of the mutations affect exons 2-13 of *GNAS*[110] and would therefore have an impact on *GNASXL* products as well, if transmitted paternally. Only a few examples of exon 1 mutations (specific to Gsα) are known.[110-112] Obesity is present in AHO patients irrespective of maternal or paternal inheritance of *GNAS* mutations. This is similar to mice carrying the *Gnas*-specific exon 1 mutation but, as discussed above, mice with a mutation of the shared exon 2 only develop obesity when the mutation is inherited maternally; paternal inheritance of the exon 2 mutation results in a lean phenotype resembling *Gnasxl* deficiency.[106,113] Thus, in mice simultaneous loss of paternally expressed *Gnasxl* and paternal-allele derived *Gnas* results in a dominant *Gnasxl* deficiency phenotype, while this does not seem to be the case in humans.

In one study, membrane preparations from adipose tissue biopsies of PHP patients showed a blunted cAMP response when stimulated with the β-adrenergic agonist isoproterenol.[114] A second study[115] diagnosed hyperinsulinemia, hyperleptinemia and reduced plasma norepinephrine levels in patients, which is reminiscent of *Gnas* m–/p+ mice. Epinephrine infusion resulted in a lower rate of lipolysis as measured through plasma concentration of free fatty acids and glycerol release, indicating adipose tissue resistance to this hormone. These findings are compatible with reduced Gsα function in adipocytes of PHP patients, but could simply be due to haploinsufficiency. This latter notion is supported by a more recent study that found biallelic expression of *GNAS* in visceral adipose tissue samples from normal individuals.[90] Similarities or potential discrepancies of *Gnas* imprinting in mouse and human deserve further attention.

It is possible that physiological deficits in newborns might not have been commonly recognized as part of AHO, since the condition is normally diagnosed later in infancy. However, early postnatal effects have been associated with other genetic defects involving the *GNAS1* locus. These mostly comprise cases of maternal UPD for chromosome 20, encompassing but also extending beyond this imprinted domain at 20q13.3. Maternal UPD for this region results in loss of paternally expressed gene products and symptoms reminiscent of *Gnasxl* deficient mice have been described, including pre- and postnatal growth retardation.[116-119] A case of constitutional deletion of the paternally inherited chromosome 20q13.3 region is associated with fetal growth retardation, postnatal hypotonia, a poor suck requiring artificial feeding and developmental delay.[120] Two further cases of small deletions of the paternal 20q chromosomal region confirm findings of pre- and postnatal growth retardation, feeding difficulties requiring artificial nutrition, hypotonia, developmental delay and also abnormal subcutaneous adipose tissue.[121] These cases, together with the description of growth-deficient children carrying a polymorphism in the *GNASXL* exon,[99] strongly suggest that human XLαs/XLN1/Alex proteins have significant roles in postnatal physiology much as their murine counterparts.

The 'Conflict Hypothesis' and Beyond

One of the most widely accepted hypotheses for the existence of genomic imprinting has been developed from kin selection theory by Haig and colleagues[122,123] and is popularly known as the 'conflict hypothesis' or the 'kinship theory of imprinting'. This hypothesis concerns the differential interests of paternally transmitted and maternally transmitted alleles in offspring over the allocation of finite resources by mothers into current offspring against the costs to lifetime reproductive success. In the actions of imprinted genes in controlling fetal growth, the predictions are obvious and the fit of the data from imprinted gene knock-outs in the mouse is remarkably good: simply put, paternally expressed imprinted genes tend to promote growth of the fetus and maternally expressed imprinted genes restrain growth.[124] For phenotypes such as postnatal metabolism and obesity in adults, such a dichotomy of maternal and paternal gene effects is far less obvious and the data contradictory. Thus, there are paternally expressed imprinted genes that when knocked-out cause obesity (*Pref1/Dlk1*) or the opposite (*Gnasxl*) and paternally expressed imprinted genes whose deficiencies reduce (*Peg3*) or enhance (*Gnasxl*) metabolic rate. Similarly, obesity is found in both PWS and AS, although these syndromes result form the loss of reciprocally imprinted genes. For a conflict-based explanation of imprinted gene action one might not expect to see such contradictions. There are several points to consider before dismissing the control of metabolism as a legitimate target for imprinting and intragenomic conflict. One could regard adult obesity as not being a particularly informative phenotype in relation to the resource allocation arguments that underlie the conflict hypothesis. It is an abnormal phenotype, which as an endpoint might obscure a variety of underlying physiological or developmental defects, so may not necessarily signal improved or impaired fitness of the genes involved. Many knock-out phenotypes are pleiotropic and their earlier effects do more obviously fit expectations of the theory. *Pref1/Dlk1* and *Peg3* both have significant effects on fetal growth in the expected direction, therefore, adult metabolic phenotypes in these cases may be epiphenomena and not the cause for primary selection of imprinting at these loci; any potential costs associated with imprinting later phenotypes are overridden by the greater advantage of imprinting their effects on fetal growth. In this respect, one could regard obesity in mice lacking expression of paternal genes as the consequence of a 'thrifty' genotype. These mutants may experience relative under-nutrition during fetal stages owing to placental insufficiency, which might result in malprogramming of neuroendocrine systems regulating energy metabolism and predisposition to develop obesity and/or diabetes later in life. Metabolic effects may therefore be indirect effects of early events.

Alternatively there may be a direct evolutionary rationale for paternally expressed genes to promote enhanced adiposity. In the context of patrilineal inbreeding that declines through the life course, it has been suggested that paternal expressed genes would favor greater maternal investment that shifts the benefits to earlier (more inbred) litters.[125] In this context, if increased adiposity is associated with precocious fertility or earlier puberty, paternally expressed genes might favor relatively enhanced adiposity.

Against these uncertainties, the *Gnas* locus is an exemplar of a role for imprinted genes in metabolism, because there are no major confounding effects on fetal growth in *Gnas* and *Gnasxl* mutants and because the oppositely imprinted gene products have contrasting effects on metabolism and seem to exist in an antagonistic relationship within a common pathway controlling metabolism. Haig has proposed an explanation for this, whereby paternally derived alleles may be selected to favor greater economy in energy expenditure in order to devote resources to individual growth, whilst maternally derived alleles are more likely to share resources (in this case warmth) with the common pool of sibs and/or half-sibs.[126] As an extension to this logic, it has been proposed that inhibition of adipocyte differentiation seen after ectopic expression of other paternally expressed imprinted genes could be an indirect effect of inhibiting development of BAT and, therefore, limiting energy expenditure in the way compatible with the conflict theory.

Through their actions on fetal growth and direct or indirect actions of adult metabolism, imprinted genes have nonetheless been considered serious contenders for genes involved in fetal programming, wherein low birth weight increases the risk of chronic adult diseases such as obesity and type 2 diabetes.[127,128] Especially so, because it has been proposed that epigenetic mechanisms could provide a link between an adverse fetal nutrition/environment and enduring or later effects on gene expression.[127,129] Because many imprinted genes can affect metabolism at several levels, programming could come about through modest but cumulative effects on a number of genes. For antagonistic gene pairs (e.g., the *Gnas* locus), where an appropriate balance of activities may be important for normal metabolism, the potential for programming may be even greater: deregulation of a common control element could result in reciprocal changes in expression that could amplify the consequences of programming. A counter argument is that some imprinting mechanisms are so robust as to be insensitive to programming. Dietary interventions in pregnant mice have been shown to lead to epigenetic changes in offspring,[130] even to apparent relaxation of imprinting[131] and we await evidence that imprinted genes are indeed targets of fetal programming.

Concluding Remarks

In conclusion, imprinted genes not only control prenatal growth, but they are also involved in the regulation of body weight in adult life and could be linked to certain types of obesity of genetic origin. Imprinted genes influence energy homeostasis at several stages; thus, they have been shown to be involved in the regulation of food intake (PWS-related genes), energy expenditure (*Gnas, Gnasxl, Peg3*), adipose tissue development (*Ndn, Peg1/Mest, Pref1/Dlk1*) and glucose homeostasis (*Atp10c, Gnasxl, Meg1/Grb10, Rasgrf1*). It seems likely that further research in knock-out mice for imprinted genes will unveil new roles for genomic imprinting in the regulation of energy homeostasis.

Acknowledgements

Work in GK's laboratory is supported by the United Kingdom's Biotechnology and Biological Sciences Research Council and the Medical Research Council. BD was supported by a research training studentship from the Medical Research Council.

References

1. Grundy SM, Brewer HB Jr, Cleeman JI et al. Definition of metabolic syndrome: Report of the National Heart, Lung and Blood Institute/American Heart Association conference on scientific issues related to definition. Circulation 2004; 109(3):433-438.
2. Ravussin E, Bogardus C. Energy balance and weight regulation: genetics versus environment. Br J Nutr 2000; 83(Suppl 1):S17-20.
3. Goldstone AP. Prader-Willi syndrome: advances in genetics, pathophysiology and treatment. Trends Endocrinol Metab 2004; 15(1):12-20.
4. Holm VA, Cassidy SB, Butler MG et al. Prader-Willi syndrome: consensus diagnostic criteria. Pediatrics 1993; 91(2):398-402.
5. Meaney FJ, Butler MG. The developing role of anthropologists in medical genetics: anthropometric assessment of the Prader-Labhart-Willi syndrome as an illustration. Med Anthropol 1989; 10(4):247-253.
6. Schwartz RS, Brunzell JD, Bierman EL. Elevated adipose tissue lipoprotein lipase in the pathogenesis of obesity in Prader-Willi syndrome. Baltimore: University Park Press; 1981.

7. Ginsberg-Fellner F, Knittle JL. Adipose tissue cellularity in the Prader-Willi Syndrome. Pediatr Res 1976; 10:409.
8. Gurr MI, Jung RT, Robinson MP et al. Adipose tissue cellularity in man: the relationship between fat cell size and number, the mass and distribution of body fat and the history of weight gain and loss. Int J Obes 1982; 6(5):419-436.
9. Nicholls RD, Knepper JL. Genome organization, function and imprinting in Prader-Willi and Angelman syndromes. Annu Rev Genomics Hum Genet 2001; 2:153-175.
10. Chai JH, Locke DP, Greally JM et al. Identification of four highly conserved genes between breakpoint hotspots BP1 and BP2 of the Prader-Willi/Angelman syndromes deletion region that have undergone evolutionary transposition mediated by flanking duplicons. Am J Hum Gene 2003; 73(4):898-925.
11. Cassidy SB. Prader-Willi syndrome. J Med Genet 1997; 34(11):917-923.
12. Lee S, Walker CL, Wevrick R. Prader-Willi syndrome transcripts are expressed in phenotypically significant regions of the developing mouse brain. Gene Expr Patterns 2003; 3(5):599-609.
13. Andrieu D, Watrin F, Niinobe M et al. Expression of the Prader-Willi gene Necdin during mouse nervous system development correlates with neuronal differentiation and p75NTR expression. Gene Expr Patterns 2003; 3(6):761-765.
14. Cummings DE, Clement K, Purnell JQ et al. Elevated plasma ghrelin levels in Prader Willi syndrome. Nat Med 2002; 8(7):643-644.
15. DelParigi A, Tschop M, Heiman ML et al. High circulating ghrelin: a potential cause for hyperphagia and obesity in prader-willi syndrome. J Clin Endocrinol Metab 2002; 87(12):5461-5464.
16. Goldstone AP, Thomas EL, Brynes AE et al. Elevated fasting plasma ghrelin in prader-willi syndrome adults is not solely explained by their reduced visceral adiposity and insulin resistance. J Clin Endocrinol Metab 2004; 89(4):1718-1726.
17. Haqq AM, Farooqi IS, O'Rahilly S et al. Serum ghrelin levels are inversely correlated with body mass index, age and insulin concentrations in normal children and are markedly increased in Prader-Willi syndrome. J Clin Endocrinol Metab 2003; 88(1):174-178.
18. Nakazato M, Murakami N, Date Y et al. A role for ghrelin in the central regulation of feeding. Nature 2001; 409(6817):194-198.
19. Tschop M, Smiley DL, Heiman ML. Ghrelin induces adiposity in rodents. Nature 2000; 407(6806):908-913.
20. Wren AM, Seal LJ, Cohen MA et al. Ghrelin enhances appetite and increases food intake in humans. J Clin Endocrinol Metab 2001; 86(12):5992.
21. Tan TM, Vanderpump M, Khoo B et al. Somatostatin infusion lowers plasma ghrelin without reducing appetite in adults with Prader-Willi syndrome. J Clin Endocrinol Metab 2004; 89(8):4162-4165.
22. Williams DL, Grill HJ, Cummings DE et al. Vagotomy dissociates short- and long-term controls of circulating ghrelin. Endocrinology 2003; 144(12):5184-5187.
23. Delrue MA, Michaud JL. Fat chance: genetic syndromes with obesity. Clin Genet 2004; 66(2):83-93.
24. Kishino T, Lalande M, Wagstaff J. UBE3A/E6-AP mutations cause Angelman syndrome. Nat Genet 1997; 15(1):70-73.
25. Matsuura T, Sutcliffe JS, Fang P et al. De novo truncating mutations in E6-AP ubiquitin-protein ligase gene (UBE3A) in Angelman syndrome. Nat Genet 1997; 15(1):74-77.
26. Lossie AC, Whitney MM, Amidon D et al. Distinct phenotypes distinguish the molecular classes of Angelman syndrome. J Med Genet 2001; 38(12):834-845.
27. Gabriel JM, Merchant M, Ohta T et al. A transgene insertion creating a heritable chromosome deletion mouse model of Prader-Willi and angelman syndromes. Proc Natl Acad Sci USA 1999; 96(16):9258-9263.
28. Cattanach BM, Barr JA, Beechey CV et al. A candidate model for Angelman syndrome in the mouse. Mamm Genome 1997; 8(7):472-478.
29. Jiang YH, Armstrong D, Albrecht U et al. Mutation of the Angelman ubiquitin ligase in mice causes increased cytoplasmic p53 and deficits of contextual learning and long-term potentiation. Neuron 1998; 21(4):799-811.
30. Miura K, Kishino T, Li E et al. Neurobehavioral and electroencephalographic abnormalities in UBE3A maternal-deficient mice. Neurobiol Dis 2002; 9(2):149-159.
31. Dhar M, Webb LS, Smith L et al. A novel ATPase on mouse chromosome 7 is a candidate gene for increased body fat. Physiol Genomics 2000; 4(1):93-100.
32. Johnson DK, Stubbs LJ, Culiat CT et al. Molecular analysis of 36 mutations at the mouse pink-eyed dilution (p) locus. Genetics 1995; 141(4):1563-1571.
33. Herzing LB, Kim SJ, Cook EH Jr. et al. The human aminophospholipid-transporting ATPase gene Atp10c maps adjacent to UBE3A and exhibits similar imprinted expression. Am J Hum Genet 2001; 68(6):1501-1505.

34. Meguro M, Kashiwagi A, Mitsuya K et al. A novel maternally expressed gene, Atp10c, encodes a putative aminophospholipid translocase associated with Angelman syndrome. Nat Genet 2001; 28(1):19-20.
35. Dhar M, Hauser L, Johnson D. An aminophospholipid translocase associated with body fat and type 2 diabetes phenotypes. Obes Res 2002; 10(7):695-702.
36. Dhar MS, Sommardahl CS, Kirkland T et al. Mice heterozygous for Atp10c, a putative amphipath, represent a novel model of obesity and type 2 diabetes. J Nutr 2004; 134(4):799-805.
37. Dhar MS, Yuan JS, Elliott SB et al. A type IV P-type ATPase affects insulin-mediated glucose uptake in adipose tissue and skeletal muscle in mice. J Nutr Biochem, 2006.
38. Lindsay RS, Kobes S, Knowler WC et al. Genome-wide linkage analysis assessing parent-of-origin effects in the inheritance of type 2 diabetes and BMI in Pima Indians. Diabetes 2001; 50(12):2850-2857.
39. Gorlova OY, Amos CI, Wang NW et al. Genetic linkage and imprinting effects on body mass index in children and young adults. Eur J Hum Genet 2003; 11(6):425-432.
40. Dong C, Li WD, Geller F et al. Possible genomic imprinting of three human obesity-related genetic loci. Am J Hum Genet 2005; 76(3):427-437.
41. Rance KA, Fustin JM, Dalgleish G et al. A paternally imprinted QTL for mature body mass on mouse chromosome 8. Mamm Genome 2005; 16(8):567-577.
42. de Koning DJ, Rattink AP, Harlizius B et al. Genome-wide scan for body composition in pigs reveals important role of imprinting. Proc Natl Acad Sci USA 2000; 97(14):7947-7950.
43. Michaud JL. The developmental program of the hypothalamus and its disorders. Clin Genet 2001; 60(4):255-263.
44. Keverne EB, Fundele R, Narasimha M et al. Genomic imprinting and the differential roles of parental genomes in brain development. Brain Res Dev Brain Res 1996; 92(1):91-100.
45. Kuroiwa Y, Kaneko-Ishino T, Kagitani F et al. Peg3 imprinted gene on proximal chromosome 7 encodes for a zinc finger protein. Nat Genet 1996; 12(2):186-190.
46. Lefebvre L, Viville S, Barton SC et al. Abnormal maternal behaviour and growth retardation associated with loss of the imprinted gene Mest. Nat Genet 1998; 20(2):163-169.
47. Kim J, Ashworth L, Branscomb E et al. The human homolog of a mouse-imprinted gene, Peg3, maps to a zinc finger gene-rich region of human chromosome 19q13.4. Genome Res 1997; 7(5):532-540.
48. Relaix F, Wei X, Li W et al. Pw1/Peg3 is a potential cell death mediator and cooperates with Siah1a in p53-mediated apoptosis. Proc Natl Acad Sci USA 2000; 97(5):2105-2110.
49. Johnson MD, Wu X, Aithmitti N et al. Peg3/Pw1 is a mediator between p53 and Bax in DNA damage-induced neuronal death. J Biol Chem 2002; 277(25):23000-23007.
50. Curley JP, Pinnock SB, Dickson SL et al. Increased body fat in mice with a targeted mutation of the paternally expressed imprinted gene Peg3. FASEB J 2005; 19(10):1302-1304.
51. Li L, Keverne EB, Aparicio SA et al. Regulation of maternal behavior and offspring growth by paternally expressed Peg3. Science 1999; 284(5412):330-333.
52. Curley JP, Barton S, Surani A et al. Coadaptation in mother and infant regulated by a paternally expressed imprinted gene. Proc Biol Sci 2004; 271(1545):1303-1309.
53. Cannon B, Nedergaard J. Brown adipose tissue: function and physiological significance. Physiol Rev 2004; 84(1):277-359.
54. Kobayashi S, Wagatsuma H, Ono R et al. Mouse Peg9/Dlk1 and human PEG9/DLK1 are paternally expressed imprinted genes closely located to the maternally expressed imprinted genes: mouse Meg3/Gtl2 and human MEG3. Genes Cells 2000; 5(12):1029-1037.
55. Takada S, Paulsen M, Tevendale M et al. Epigenetic analysis of the Dlk1-Gtl2 imprinted domain on mouse chromosome 12: implications for imprinting control from comparison with Igf2-H19. Hum Mol Genet 2002; 11(1):77-86.
56. Wylie AA, Murphy SK, Orton TC et al. Novel imprinted DLK1/GTL2 domain on human chromosome 14 contains motifs that mimic those implicated in IGF2/H19 regulation. Genome Res 2000; 10(11):1711-1718.
57. Laborda J, Sausville EA, Hoffman T et al. Dlk, a putative mammalian homeotic gene differentially expressed in small cell lung carcinoma and neuroendocrine tumor cell line. J Biol Chem 1993; 268(6):3817-3820.
58. Smas CM, Sul HS. Pref-1, a protein containing EGF-like repeats, inhibits adipocyte differentiation. Cell 1993; 73(4):725-734.
59. Schmidt JV, Matteson PG, Jones BK et al. The Dlk1 and Gtl2 genes are linked and reciprocally imprinted. Genes Dev 2000; 14(16):1997-2002.
60. Larsen JB, Jensen CH, Schroder HD et al. Fetal antigen 1 and growth hormone in pituitary somatotroph cells. Lancet 1996; 347(8995):191.
61. Jensen CH, Teisner B, Hojrup P et al. Studies on the isolation, structural analysis and tissue localization of fetal antigen 1 and its relation to a human adrenal-specific cDNA, pG2. Hum Reprod 1993; 8(4):635-641.

62. Moon YS, Smas CM, Lee K et al. Mice lacking paternally expressed Pref-1/Dlk1 display growth retardation and accelerated adiposity. Mol Cell Biol 2002; 22(15):5585-5592.
63. Smas CM, Chen L, Sul HS. Cleavage of membrane-associated pref-1 generates a soluble inhibitor of adipocyte differentiation. Mol Cell Biol 1997; 17(2):977-988.
64. Smas CM, Sul HS. Molecular mechanisms of adipocyte differentiation and inhibitory action of pref-1. Crit Rev Eukaryot Gene Expr 1997; 7(4):281-298.
65. Kaneko-Ishino T, Kuroiwa Y, Miyoshi N et al. Peg1/Mest imprinted gene on chromosome 6 identified by cDNA subtraction hybridization. Nat Genet1995; 11(1):52-59.
66. Kobayashi S, Kohda T, Miyoshi N et al. Human PEG1/MEST, an imprinted gene on chromosome 7. Hum Mol Genet 1997; 6(5):781-786.
67. Reule M, Krause R, Hemberger M et al. Analysis of Peg1/Mest imprinting in the mouse. Dev Genes Evol 1998; 208(3):161-163.
68. Takahashi M, Kamei Y, Ezaki O. Mest/Peg1 imprinted gene enlarges adipocytes and is a marker of adipocyte size. Am J Physiol Endocrinol Metab 2005; 288(1):E117-124.
69. Shi W, Lefebvre L, Yu Y et al. Loss-of-imprinting of Peg1 in mouse interspecies hybrids is correlated with altered growth. Genesis 2004; 39(1):65-72.
70. Koza RA, Nikonova L, Hogan J et al. Changes in gene expression foreshadow diet-induced obesity in genetically identical mice. PLoS Genet 2006; 2(5):e81.
71. Taniura H, Taniguchi N, Hara M et al. Necdin, a postmitotic neuron-specific growth suppressor, interacts with viral transforming proteins and cellular transcription factor E2F1. J Biol Chem 1998; 273(2):720-728.
72. Gerard M, Hernandez L, Wevrick R et al. Disruption of the mouse necdin gene results in early post natal lethality. Nat Genet 1999; 23(2):199-202.
73. Boeuf S, Klingenspor M, Van Hal NL et al. Differential gene expression in white and brown preadipocytes. Physiol Genomics 2001; 7(1):15-25.
74. Tseng YH, Butte AJ, Kokkotou E et al. Prediction of preadipocyte differentiation by gene expression reveals role of insulin receptor substrates and necdin. Nat Cell Biol 2005; 7(6):601-611.
75. Tsai TF, Jiang YH, Bressler J et al. Paternal deletion from Snrpn to UBE3A in the mouse causes hypotonia, growth retardation and partial lethality and provides evidence for a gene contributing to Prader-Willi syndrome. Hum Mol Genet 1999; 8(8):1357-1364.
76. DeChiara TM, Efstratiadis A, Robertson EJ. A growth-deficiency phenotype in heterozygous mice carrying an insulin-like growth factor II gene disrupted by targeting. Nature 1990; 345(6270):78-80.
77. Jones BK, Levorse J, Tilghman SM. Deletion of a nuclease-sensitive region between the Igf2 and H19 genes leads to Igf2 misregulation and increased adiposity. Hum Mol Genet 2001; 10(8):807-814.
78. Weinstein LS, Yu S, Warner DR et al. Endocrine manifestations of stimulatory G protein alpha-subunit mutations and the role of genomic imprinting. Endocr Rev 2001; 22(5):675-705.
79. Wettschureck N, Offermanns S. Mammalian G proteins and their cell type specific functions. Physiol Rev 2005; 85(4):1159-1204.
80. Yu S, Yu D, Lee E et al. Variable and tissue-specific hormone resistance in heterotrimeric Gs protein alpha-subunit (Gsalpha) knockout mice is due to tissue-specific imprinting of the gsalpha gene. Proc Natl Acad Sci USA 1998; 95(15):8715-8720.
81. Williamson CM, Ball ST, Nottingham WT et al. A cis-acting control region is required exclusively for the tissue-specific imprinting of Gnas. Nat Genet 2004; 36(8):894-899.
82. Chen M, Gavrilova O, Liu J et al. Alternative Gnas gene products have opposite effects on glucose and lipid metabolism. Proc Natl Acad Sci USA 2005; 102(20):7386-7391.
83. Liu J, Chen M, Deng C et al. Identification of the control region for tissue-specific imprinting of the stimulatory G protein alpha-subunit. Proc Natl Acad Sci USA 2005; 102(15):5513-5518.
84. Germain-Lee EL, Schwindinger W, Crane JL et al. A mouse model of albright hereditary osteodystrophy generated by targeted disruption of exon 1 of the Gnas gene. Endocrinology 2005; 146(11):4697-4709.
85. Hayward BE, Barlier A, Korbonits M et al. Imprinting of the G(s)alpha gene GNAS1 in the pathogenesis of acromegaly. J Clin Invest 2001; 107(6):R31-36.
86. Germain-Lee EL, Ding CL, Deng Z et al. Paternal imprinting of Galpha(s) in the human thyroid as the basis of TSH resistance in pseudohypoparathyroidism type 1a. Biochem Biophys Res Commun 2002; 296(1):67-72.
87. Germain-Lee EL, Groman J, Crane JL et al. Growth hormone deficiency in pseudohypoparathyroidism type 1a: another manifestation of multihormone resistance. J Clin Endocrinol Metab 2003; 88(9):4059-4069.
88. Liu J, Erlichman B, Weinstein LS. The stimulatory G protein alpha-subunit Gs alpha is imprinted in human thyroid glands: implications for thyroid function in pseudohypoparathyroidism types 1A and 1B. J Clin Endocrinol Metab 2003; 88(9):4336-4341.

89. Mantovani G, Ballare E, Giammona E et al. The gsalpha gene: predominant maternal origin of transcription in human thyroid gland and gonads. J Clin Endocrinol Metab 2002; 87(10):4736-4740.
90. Mantovani G, Bondioni S, Locatelli M et al. Biallelic expression of the Gsalpha gene in human bone and adipose tissue. J Clin Endocrinol Metab 2004; 89(12):6316-6319.
91. Kehlenbach RH, Matthey J, Huttner WB. XL alpha s is a new type of G protein. Nature 1994; 372(6508):804-809.
92. Klemke M, Pasolli HA, Kehlenbach RH et al. Characterization of the extra-large G protein alpha-subunit XLalphas. II. Signal transduction properties. J Biol Chem 2000; 275(43):33633-33640.
93. Ugur O, Jones TL. A proline-rich region and nearby cysteine residues target XLalphas to the Golgi complex region. Mol Biol Cell 2000; 11(4):1421-1432.
94. Bastepe M, Gunes Y, Perez-Villamil B et al. Receptor-mediated adenylyl cyclase activation through XLalpha(s), the extra-large variant of the stimulatory G protein alpha-subunit. Mol Endocrinol 2002; 16(8):1912-1919.
95. Linglart A, Mahon MJ, Kerachian MA et al. Coding Gnas mutations leading to hormone resistance impair in vitro agonist- and cholera toxin-induced adenosine cyclic 3',5'-monophosphate formation mediated by human XLalphas. Endocrinology 2006; 147(5):2253-2262.
96. Pasolli HA, Klemke M, Kehlenbach RH et al. Characterization of the extra-large G protein alpha-subunit XLalphas. I. Tissue distribution and subcellular localization. J Biol Chem 2000; 275(43):33622-33632.
97. Plagge A, Gordon E, Dean W et al. The imprinted signaling protein XL alpha s is required for postnatal adaptation to feeding. Nat Genet 2004; 36(8):818-826.
98. Nekrutenko A, Wadhawan S, Goetting-Minesky P et al. Oscillating evolution of a mammalian locus with overlapping reading frames: an XLalphas/ALEX relay. PLoS Genet 2005; 1(2):e18.
99. Freson K, Jaeken J, Van Helvoirt M et al. Functional polymorphisms in the paternally expressed XLalphas and its cofactor ALEX decrease their mutual interaction and enhance receptor-mediated cAMP formation. Hum Mol Genet 2003; 12(10):1121-1130.
100. Freson K, Hoylaerts MF, Jaeken J et al. Genetic variation of the extra-large stimulatory G protein alpha-subunit leads to Gs hyperfunction in platelets and is a risk factor for bleeding. Thromb Haemost 2001; 86(3):733-738.
101. Ischia R, Lovisetti-Scamihorn P, Hogue-Angeletti R et al. Molecular cloning and characterization of NESP55, a novel chromogranin-like precursor of a peptide with 5-HT1B receptor antagonist activity. J Biol Chem 1997; 272(17):11657-11662.
102. Plagge A, Isles AR, Gordon E et al. Imprinted Nesp55 influences behavioral reactivity to novel environments. Mol Cell Biol 2005; 25(8):3019-3026.
103. Plagge A, Kelsey G. Imprinting the Gnas locus. Cytogenet Genome Res 2006; 113(1-4):178-187.
104. Weinstein LS, Liu J, Sakamoto A et al. Minireview: Gnas: normal and abnormal functions. Endocrinology 2004; 145(12):5459-5464.
105. Cattanach BM, Peters J, Ball S et al. Two imprinted gene mutations: three phenotypes. Hum Mol Genet 2000; 9(15):2263-2273.
106. Yu S, Gavrilova O, Chen H et al. Paternal versus maternal transmission of a stimulatory G-protein alpha subunit knockout produces opposite effects on energy metabolism. J Clin Invest 2000; 105(5):615-623.
107. Chen M, Gavrilova O, Zhao WQ et al. Increased glucose tolerance and reduced adiposity in the absence of fasting hypoglycemia in mice with liver-specific Gs alpha deficiency. J Clin Invest 2005; 115(11):3217-3227.
108. Xie T, Plagge A, Gavrilova O et al. The alternative stimulatory G protein alpha-subunit XLalpha s is a critical regulator of energy and glucose metabolism and sympathetic nerve activity in adult mice. J Biol Chem, 2006.
109. Yu S, Castle A, Chen M et al. Increased insulin sensitivity in Gsalpha knockout mice. J Biol Chem 2001; 276(23):19994-19998.
110. Aldred MA, Trembath RC. Activating and inactivating mutations in the human GNAS1 gene. Hum Mutat 2000; 16(3):183-189.
111. Fischer JA, Egert F, Werder E et al. An inherited mutation associated with functional deficiency of the alpha-subunit of the guanine nucleotide-binding protein Gs in pseudo- and pseudopseudohypoparathyroidism. J Clin Endocrinol Metab 1998; 83(3):935-938.
112. Patten JL, Johns DR, Valle D et al. Mutation in the gene encoding the stimulatory G protein of adenylate cyclase in Albright's hereditary osteodystrophy. N Engl J Med 1990; 322(20):1412-1419.
113. Chen M, Haluzik M, Wolf NJ et al. Increased insulin sensitivity in paternal Gnas knockout mice is associated with increased lipid clearance. Endocrinology 2004; 145(9):4094-4102.
114. Kaartinen JM, Kaar ML, Ohisalo JJ. Defective stimulation of adipocyte adenylate cyclase, blunted lipolysis and obesity in pseudohypoparathyroidism 1a. Pediatr Res 1994; 35(5):594-597.

115. Carel JC, Le Stunff C, Condamine L et al. Resistance to the lipolytic action of epinephrine: a new feature of protein Gs deficiency. J Clin Endocrinol Metab 1999; 84(11):4127-4131.
116. Chudoba I, Franke Y, Senger G et al. Maternal UPD 20 in a hyperactive child with severe growth retardation. Eur J Hum Genet 1999; 7(5):533-540.
117. Eggermann T, Mergenthaler S, Eggermann K et al. Identification of interstitial maternal uniparental disomy (UPD) (14) and complete maternal UPD(20) in a cohort of growth retarded patients. J Med Genet 2001; 38(2):86-89.
118. Salafsky IS, MacGregor SN, Claussen U et al. Maternal UPD 20 in an infant from a pregnancy with mosaic trisomy 20. Prenat Diagn 2001; 21(10):860-863.
119. Velissariou V, Antoniadi T, Gyftodimou J et al. Maternal uniparental isodisomy 20 in a foetus with trisomy 20 mosaicism: clinical, cytogenetic and molecular analysis. Eur J Hum Genet 2002; 10(11):694-698.
120. Aldred MA, Aftimos S, Hall C et al. Constitutional deletion of chromosome 20q in two patients affected with albright hereditary osteodystrophy. Am J Med Genet 2002; 113(2):167-172.
121. Genevieve D, Sanlaville D, Faivre L et al. Paternal deletion of the Gnas imprinted locus (including Gnasxl) in two girls presenting with severe pre and post natal growth retardation and intractable feeding difficulties. Eur J Hum Genet 2005; 13(9):1033-1039.
122. Haig D, Westoby M. Parent-specific gene expression and the triploid endosperm. Am Nat 1989; 134:147-155.
123. Moore T, Haig D. Genomic imprinting in mammalian development: a parental tug-of-war. Trends Genet 1991; 7(2):45-49.
124. Wilkins JF, Haig D. What good is genomic imprinting: the function of parent-specific gene expression. Nat Rev Genet 2003; 4(5):359-368.
125. Wilkins JF, Haig D. Inbreeding, maternal care and genomic imprinting. J Theor Biol 2003; 221(4):559-564.
126. Haig D. Genomic imprinting and kinship: how good is the evidence? Annu Rev Genet 2004; 38:553-585.
127. Gallou-Kabani C, Junien C. Nutritional epigenomics of metabolic syndrome: new perspective against the epidemic. Diabetes 2005; 54(7):1899-1906.
128. Constancia M, Kelsey G, Reik W. Resourceful imprinting. Nature 2004; 432(7013):53-57.
129. Waterland RA, Jirtle RL. Early nutrition, epigenetic changes at transposons and imprinted genes and enhanced susceptibility to adult chronic diseases. Nutrition 2004; 20(1):63-68.
130. Waterland RA, Jirtle RL. Transposable elements: targets for early nutritional effects on epigenetic gene regulation. Mol Cell Biol 2003; 23(15):5293-5300.
131. Waterland RA, Lin JR, Smith CA et al. Post weaning diet affects genomic imprinting at the insulin-like growth factor 2 (Igf2) locus. Hum Mol Genet 2006; 15(5):705-716.
132. Dhar MS, Hauser LJ, Nicholls RD et al. Physical mapping of the pink-eyed dilution complex in mouse chromosome 7 shows that Atp10c is the only transcript between Gabrb3 and UBE3A. DNA Seq 2004; 15(4):306-309.
133. Font de Mora J, Esteban LM, Burks DJ et al. Ras-GRF1 signaling is required for normal beta-cell development and glucose homeostasis. EMBO J 2003; 22(12):3039-3049.
134. Shiura H, Miyoshi N, Konishi A et al. Meg1/Grb10 overexpression causes postnatal growth retardation and insulin resistance via negative modulation of the IGF1R and IR cascades. Biochem Biophys Res Commun 2005; 329(3):909-916.
135. Hernandez A, Martinez ME, Fiering S et al. Type 3 deiodinase is critical for the maturation and function of the thyroid axis. J Clin Invest 2006; 116(2):476-484.
136. Lee K, Villena JA, Moon YS et al. Inhibition of adipogenesis and development of glucose intolerance by soluble preadipocyte factor-1 (Pref-1). J Clin Invest 2003; 111(4):453-461.
137. Cattanach BM, Kirk M. Differential activity of maternally and paternally derived chromosome regions in mice. Nature 1985; 315(6019):496-498.
138. Williamson CM, Beechey CV, Papworth D et al. Imprinting of distal mouse chromosome 2 is associated with phenotypic anomalies in utero. Genet Res 1998; 72(3):255-265.

CHAPTER 5

What Are Imprinted Genes Doing in the Brain?

William Davies,* Anthony R. Isles, Trevor Humby and Lawrence S. Wilkinson

Abstract

As evidence for the existence of brain-expressed imprinted genes accumulates, we need to address exactly what they are doing in this tissue, especially in terms of organisational themes and the major challenges posed by reconciling imprinted gene action in brain with current evolutionary theories attempting to explain the origin and maintenance of genomic imprinting. We are at the beginning of this endeavor and much work remains to be done but already it is clear that imprinted genes have the potential to influence diverse behavioral processes via multiple brain mechanisms. There are also grounds to believe that imprinting may contribute to risk of mental and neurological disease. As well as being a source of basic information about imprinted genes in the brain (e.g., via the newly established website, www.bgg.cardiff.ac.uk/imprinted_tables/index.html), we have used this chapter to identify and focus on a number of key questions. How are brain-expressed imprinted genes organised at the molecular and cellular levels? To what extent does imprinted action depend on neurodevelopmental mechanisms? Do imprinted gene effects interact with other epigenetic influences, especially early on in life? Are imprinted effects on adult behaviors adaptive or just epiphenomena? If they are adaptive, what areas of brain function and behavior might be sensitive to imprinted effects? These are big questions and, as shall become apparent, we need much more data, arising from interactions between behavioral neuroscientists, molecular biologists and evolutionary theorists, if we are to begin to answer them.

Imprinted Genes and the Brain

Imprinted genes, in contrast to most mammalian genes, are monoallelically expressed in a parent-of-origin dependent manner.[1] Early evidence from human and animal studies suggested that they were likely to play a fundamental role in basic growth and development, but recent work has suggested that in some cases they may also mediate more subtle effects on ongoing physiological processes. Imprinted genes are expressed in a wide range of tissues, but are particularly highly expressed in the placenta[2] and the brain.[3] Although much work remains to be done, there is now convincing and convergent functional evidence from a variety of sources (summarized below) that imprinted genes play an important role in the development and/or the ongoing function of the brain and that imprinted gene dysfunction may predispose to several common neurological/neuropsychiatric disorders.[4-6] Addressing the extent to which imprinted genes contribute to normal brain development and function and therefore the extent to which imprinted gene malfunction may be implicated in brain abnormalities, will constitute an important focus of research over the

---

*Corresponding Author: William Davies—Department of Psychological Medicine, University of Cardiff, Henry Wellcome Building, Heath Park, Cardiff, Wales CF14 4XN, U.K. Email: daviesw4@cardiff.ac.uk

Genomic Imprinting, edited by Jon F. Wilkins. ©2008 Landes Bioscience and Springer Science+Business Media.

coming years. In order to facilitate these objectives, we have recently established a freely accessible and updatable database listing all brain-expressed imprinted genes, their expression patterns, their putative functions and information on their imprinting status in other tissues (www.bgg.cardiff.ac.uk/imprinted_tables/index.html).

Summary Evidence for a Role for Imprinted Genes in Brain Function

- Mouse models in which the dosage of imprinted genes has been reduced (e.g., by deletion of the gene) or enhanced (e.g., by uniparental disomy) commonly show brain and behavioral phenotypes (reviewed in ref. 7).
- Cells containing solely paternally or maternally inherited diploid genomes localize to different brain regions and appear to affect brain growth in mice[8] (and see later).
- Some behavioral phenotypes in mice (e.g., urinary odour preference) may be subject to parent-of-origin effects consistent with the influence of underlying imprinted genes.[9,10]
- In humans, cytogenetic disruptions of imprinted gene-rich regions may be associated with aberrant neurobiology e.g., 15q11-q13 abnormalities lead to the neurobehavioral disorders Prader-Willi and Angelman syndromes (PWS and AS respectively),[11] and maternal duplications of this region may also be related to autistic phenotypes.[5]
- Some linkage and association findings are sensitive to the parental origin of the region of interest, perhaps indicating the influence of underlying imprinted genes. For example, looking at a quantitative behavioral trait (handedness), Francks et al only found significant linkage to 2p12-q11 in the case of paternal identity-by-descent sharing.[12] Similarly, in a linkage screen for autism genes, findings on 7q were parent-of-origin dependent.[13]
- Certain neuropsychiatric and neurological disorders (e.g., Tourette's syndrome, panic disorder, multiple sclerosis, Alzheimer's disease, schizophrenia and bipolar disorder) appear to be preferentially transmitted from one parent, again potentially suggestive of underlying imprinted genes, though other mechanisms could also explain this bias.[5]

Characteristics of Brain-Expressed Imprinted Genes

Given our current limited state of knowledge of brain-expressed imprinted genes, it is difficult, at present, to identify any common themes uniting imprinted gene action. Certainly, a cursory look at the data available at the molecular level seems to indicate a multitude of functions, from neurotransmitter receptor subunits to proteins involved in chromatin modification.[3] Furthermore, imprinting in the brain is a complex and spatiotemporally dynamic process (even more so than in other organs given the heterogeneity of the tissue) and these factors constitute a major challenge when examining imprinting at the cellular level. There is however, evidence consistent with the existence of some degree of patterning with respect to cell-type specific imprinting. Cell culture studies have revealed that murine *Ube3a* sense transcripts are maternally expressed in neurons and biallelically expressed in glia, whereas the paternally expressed antisense transcript, *Ube3as*, is expressed only in neurons.[14] In contrast, *Igf2r* is preferentially maternally expressed (with its antisense transcript *Air* paternally expressed) in glial cell cultures, but in neurons *Igf2r* is biallelically expressed and *Air* is not expressed.[15] The significance of this cell-specific imprinting remains to be determined, but it is certainly possible that, for some genes at least, there is some functional correlation between their imprinted status and their roles in different cell types.

In terms of their spatial expression patterns, most brain-expressed imprinted genes seem to be expressed and imprinted in at least one other organ; however, there are several exceptions to this rule. Studying these exceptions may give important insights into the precise role of imprinting in the brain. For example, the imprinted *Nnat* gene (encoding neuronatin) appears to be expressed in a relatively neural-specific manner,[16] whilst the genes *Zim1* (encoding a zinc finger protein)[17] and *Ppp1r9a* (encoding neurabin) are imprinted in other tissues but biallelically expressed in the brain. Conversely, *UBE3A* is biallelically expressed everywhere but the brain,[18] whilst its murine homologue *Ube3a* is biallelically expressed throughout most of the brain, but is imprinted in specific brain regions (olfactory bulb, hippocampus and Purkinje cells of the cerebellum);[19] *Igf2* may also only show imprinted expression in highly specific brain regions.[20] In general, brain-expressed genes seem to be persistently imprinted (although overall expression levels of several imprinted genes including *Xlr3b*,[21] *Ndn*, *Mkrn3* and *Magel2* (D. Relkovic, L.S. Wilkinson and A.R. Isles, unpublished data) decline markedly throughout development, perhaps consistent with them exerting their greatest effect during embryogenesis). However, for a number of these genes, imprinting has only been demonstrated at a single time-point, so the generality of this conclusion remains to be tested. There are also exceptions to this pattern of imprinted expression occurring predominantly during early development. *Commd1/Murr1* shows strong maternal expression in adult brain, but biallelic, or weak maternal expression in embryonic and neonatal brain.[22]

Imprinted Gene Effects on Brain Development

Seminal mouse studies in the mid 1990s revealed that imprinted genes are likely to contribute significantly to brain development and also indicated potentially dissociable (and antagonistic) influences of paternally and maternally expressed genes on this process. Briefly, chimeric mice were created which contained either a mixture of androgenetic (Ag) (containing two paternal genomes but no maternal genome) and normal cells, or a mixture of parthenogenetic (Pg) (containing two maternal genomes but no paternal genome) and normal cells. 'Pg chimeras' displayed relatively large brain:body size ratios, whilst 'Ag chimeras' displayed relatively small brain:body size ratios, implying that one or more imprinted genes have profound effects on brain size.[8] Specifically, the data seem to indicate that the overall effect of maternally expressed genes is to enhance brain size, whilst the combined effect of paternally expressed genes is to limit brain growth. More detailed neuroanatomical studies on the Ag and Pg chimeras will enable us to decipher why their brains differ in size (and hence which processes the underlying imprinted genes may be affecting). Increased brain size in Pg chimeras could theoretically be due to increased proliferation of neuronal (or glial) cells, an increase in neuronal (or glial) size perhaps related to aberrant pruning, or a decrease in cell death/apoptosis. Interestingly, the distribution of the Pg and Ag cells in the two types of chimera was reciprocal, with Pg cells contributing mainly to the neocortex and Ag cells contributing more to the hypothalamic, septal and pre-optic areas. Again, it is feasible that these effects represent either the combined effects of many paternally and maternally expressed imprinted genes, or the actions of one or two imprinted genes of major effect. If the former is the case, we may expect maternally expressed imprinted genes to be disproportionately expressed in neocortical regions and paternally expressed imprinted genes to be disproportionately expressed in hypothalamic and septal regions. However, based on our current knowledge this conclusion does not appear to be supported.[3] For a further discussion of these chimeric mouse experiments, see the chapters by Goos and Ragsdale and by Frontera et al.

Further evidence for an important role of imprinted genes in neurodevelopment has come from studies of mutant mice and humans with imprinting disorders such as AS and PWS. Currently such studies are rare, but as the number of imprinted genes discovered and knockout mice created continues to rise (and as the significance of imprinted gene function to neurobiological processes begins to be realized) more data will accumulate. In mice, the deletion of *Ndn*, a candidate gene for PWS, results in morphological abnormalities in axonal outgrowth and fasciculation in several regions of the nervous system during embryogenesis.[23] Necdin-deficient mice exhibit a reduction in both oxytocin-producing and luteinizing hormone-releasing hormone (LHRH)-producing neurons in adult hypothalamus,[24] augmented apoptosis in the sensory ganglia and a reduction in the numbers of substance-P containing neurons.[25] These neuroanatomical abnormalities may explain the characteristic PWS behavioral profile (described later). Deletion of *Peg3* similarly results in reduced numbers of oxytocin-producing neurons in the hypothalamus of adult female mice.[26] Deletion of the AS candidate gene *Ube3a* does not affect gross neuroanatomy but does lead to accumulation of cytoplasmic p53 and deficits in long term potentiation[27,28] whilst abnormal cerebellar folding was observed in mice disomic for distal chromosome 2, including *Nnat*.[29] In some instances, such as in deletion of *Gnasxl*, *RasGrf1* and *Nesp*, no gross effects on brain morphology have been observed; however subtle effects on fundamental processes such as neuronal outgrowth in the case of *RasGrf1* knockouts[30] and vesicular release in the case of *Nesp* knockouts[31] cannot be discounted. More comprehensive neuroanatomical studies using imprinted gene knockouts will enable us to identify such abnormalities.

With respect to human work, imaging studies in PWS subjects have revealed abnormalities in pituitary, cortical and brainstem development, whilst studies in AS subjects have shown cerebellar and Sylvian fissure abnormalities, with ventricular enlargement and hypoperfusion of some brain regions.[3] In studies such as these, where subjects do not have highly discrete genetic lesions, caution must be applied during interpretation since any neuroanatomical phenotype could simply result from the aberrant expression of non-imprinted brain-expressed genes rather than from imprinted gene effects per se.

Imprinted Gene Effects on Behavior

Parallel work in humans and mice has suggested that imprinted genes may not only affect brain development but may also impact upon a wide range of behavioral phenotypes. Explicit imprinted conditions such as PWS and AS are associated with distinct behavioral profiles. PWS is characterised by early hypotonia followed by a compulsive desire to eat (probably reflecting impaired satiety mechanisms[32]) and is associated with mild mental retardation, insensitivity to pain, tantrums, obsessive tendencies, skin picking, unusual skill with jigsaws and in some cases psychosis.[33,34] AS is characterised by, mental retardation, ataxia, what has been termed a 'happy' disposition and repetitive or stereotyped behaviors.[35] These behavioral phenotypes imply that imprinted genes may affect both primary-motivated behaviors and higher-level cognitive functions; however, again there is the possibility of contributions from abnormal expression of non-imprinted brain-expressed genes. There is also other evidence for imprinted gene effects on behavior in man related to the sex chromosomes. For example, the behavioral profile of Turner's syndrome (TS) subjects (who possess a single X chromosome) depends upon the parental origin of this chromosome, suggesting an influence of one or more, as yet unidentified, X-linked imprinted genes. TS subjects inheriting their X chromosome maternally ($45,X^mO$) display impaired social cognition (specifically behavioral flexibility) and are significantly more vulnerable to autism than subjects inheriting their single X chromosome paternally ($45,X^PO$);[36] however, the former group demonstrate superior performance than the latter group on a task assaying visuospatial memory.[37] As X-linked imprinted genes may be expressed in a sexually dimorphic manner and may influence sex-limited behavioral traits and sex-specific vulnerability to certain mental disorders,[38,39] the identification of such genes in man is an important future goal. Clues as to the identity of human X-linked imprinted genes may be gained from the recent identification of novel X-linked imprinted genes in mice.[21,40] Other aspects of the phenotypes associated with AS, PWS and TS are discussed in the chapters by Frontera et al and by Goos and Ragsdale.

In mice, the first behavioral phenotype to be associated with imprinted gene activity was observed in neonatal mice disomic for chromosome 2; mice with paternal uniparental disomy appeared hyperkinetic whilst mice with maternal uniparental disomy were hypokinetic.[41] Since that study, more sophisticated behavioral analyses have been performed on a number of imprinted gene knockout mice. Behavioral work so far has mainly been directed by our limited knowledge of the expression patterns and functions of the disrupted imprinted genes; this ascertainment bias means that only a small proportion of the repertoire of behaviors that imprinted genes influence is likely to have been uncovered as yet. Thus, it is still too early to say whether the deletion of paternally and maternally expressed genes leads to dissociable and opposite phenotypes, a key question. Ideally, a given mutant should undergo a comprehensive battery of behavioral tests assaying sensorimotor, 'emotional' and cognitive functions to reveal any interesting phenotypes not predicted a priori. Yet even now it is clear that imprinted genes may influence a wide range of murine behaviors, from primary motivated feeding behaviors to higher level cognitive processes. In neonates, deletion of the paternally expressed *Gnasxl* leads to impairments in suckling, consistent with expression of this gene in regions related to innervation of the tongue and jaw muscles.[42] In adult mice, deletion of the paternally expressed genes *Peg1/Mest* and *Peg3* is associated with impaired mothering,[26,43] presumably a consequence of abnormal hypothalamic development. Deletion of *Ube3a* has resulted in deficits of context-dependent memory[27] whilst deletion of *RasGrf1* has been shown to lead to impairments in memory consolidation[44] and possibly long term depression.[45] Deletion of *Ndn* recapitulates some of the behavioral facies seen in PWS including improved spatial learning and memory and skin scraping.[24] Finally, deletion of the maternally expressed *Nesp* gene results in abnormal reactivity to novel environments in adult animals.[46] Importantly, for several of these knockout mice, the behavioral effects in adults are independent of gross effects of the gene on development (see later).

Through What Mechanisms Might Imprinted Genes Affect (Adult) Behavior?

As many imprinted genes are expressed in the brain and placenta during embryogenesis and many remain expressed in adult brain, a number of possibilities exist with regard to how they may affect adult brain function (and hence how they may affect vulnerability to adult psychiatric disease when aberrantly expressed). The expression of some imprinted genes during neurodevelopment (e.g., *Ndn*) may have direct and major, effects on neuronal growth and pruning, axonal sprouting and interconnections that take place during this critical period of time, with important consequences for brain functionality and connectivity. A recent bioinformatic screen in mice has predicted imprinting of two crucial genes in brain development, *Bdnf* and *Gdnf*;[47] whether these genes are actually imprinted in mice, (and/or humans), remains to be tested, but the functional ramifications of imprinting of these genes would be far-reaching.

However, several imprinted genes that are not expressed in brain may also theoretically affect neural function in an indirect manner. For example, imprinted genes such as *Igf2* and *Slc38a4* seem to have a major role in governing transfer of essential nutrients (e.g., glucose, amino acids) across the placental membranes,[48] and too much or too little of these resources could lead to many downstream effects in the future, abnormal brain development being one of them. Moreover, these changes may not show directly in the offspring but may lead to an initially silent 'programming' that only becomes effective in adulthood. Such latent programmes may become activated in adulthood, e.g., as a result of a stressor, the idea behind Barker's hypothesis; 'the developmental origins of adult disease'.[49] Conditions such as type-2 diabetes, hypertension and osteoporosis have been the major focus of such work but certain triggers may in some circumstances result in an increased vulnerability to mental and behavioral problems. Such effects may be particularly apparent if an individual, when born small (e.g., as a consequence of placental insufficiency), experiences a period of 'catch-up' growth.[50,51] Imprinted genes may also influence adult phenotypes at another key developmental stage, the preweaning environment and there is at least one imprinted gene, *RasGrf1*, that does not affect gross embryonic development (knockout mice were of equivalent weights to their wildtype littermates at birth), but that may affect perinatal growth.[52] Therefore, in addition to affecting neurodevelopment directly (the gene is expressed throughout the brain) *RasGrf1* may also affect brain development indirectly, via effects on preweaning growth and development. It is well established that early life mother-pup interactions can profoundly influence the development and function of the nervous system, as highlighted, for example, by studies examining the neurobiological sequelae of maternal deprivation in rodents.[53] Thus, imprinted genes such as *Peg1* and *Peg3*, which influence behavior of mothers towards their offspring, may shape the future behavior of their offspring via this epigenetic route. Recent exciting work by Michael Meaney's group, in which they showed that the epigenetic status of the glucocorticoid receptor and estrogen receptor genes and neuronal survivability in the hippocampus of pups depends upon the level of maternal care (specifically licking),[54-56] has highlighted the possibility of an imprinted gene-dependent 'loop' i.e., imprinted genes may influence certain types of behavior (such as maternal licking and grooming), which in turn elicit effects on the expression of genes particularly sensitive to epigenetic perturbation in their offspring, potentially including imprinted genes themselves.

The ways in which imprinted genes affect brain development and function during embryogenesis and the perinatal period are, as can be appreciated from the above, likely to be complex and inter-dependent. Additionally, the fact that many imprinted genes are expressed into adulthood implies that imprinted gene function may be directly relevant to adult brain function per se. The most convincing explanation for the evolutionary origins of imprinting, the conflict theory, posits that imprinting has arisen as a consequence of the differential interests of the two parents with regard to provisioning of their common offspring (see Chapter by Moore and Mills). A recent adaptation of the original conflict theory, the kinship theory, predicts that asymmetries of relatedness within a social group (e.g., when there is sex-biased dispersal) may provide a route by which intragenomic conflict impacting upon behavioral functioning could arise.[57,58] This idea is important on two levels; firstly it suggests that imprinted genes may influence adult brain function; second it points to (in the broadest sense) the behavioral substrates that may be influenced, namely social interactions within the groups.[58,59]

Imprinted Genes in the Adult Brain

Given the preponderance of data showing that many imprinted genes have gross effects on growth (whether prenatal or peri/postnatal), to what extent may adult imprinted brain functions be ascribed to epiphenomena? Is the primary (and adaptive) function of all adult brain expressed imprinted genes simply growth modulation?[60] There is some, admittedly limited, data that address this issue. Firstly, many genes show continued imprinting in the adult brain and some, like *Ube3a*, show imprinting in specific and discrete regions of the brain. Given that monoallelic expression of a gene may be detrimental, insofar as the advantages of diploidy are effectively lost,[61] this implies some degree of specific functionality in the imprinting status of a gene persistently expressed in the adult brain. (See also the chapter by Ubeda and Wilkins.) Additionally, as mentioned previously, the gene *Commd/Murr1* only appears to be fully imprinted in the adult brain.[22] Thus, we might speculate that imprinting may not only be important for early life processes, but that it may also be functionally important for so-called 'online' adult brain functions. Altering the dosage of such a gene in mouse mutants, together with further investigations into its neural function will allow us to determine whether its imprinting in the adult is likely to be of any adaptive significance, or whether it is merely a redundant side-effect. Recently we, in collaboration with the group of Gavin Kelsey, have shown that the maternally expressed gene *Nesp*, part of the *Gnas* cluster, influences adult behavior. Mice carrying a maternally derived null allele of *Nesp* demonstrated altered reactivity to a novel environment as adults, with no obvious concomitant effects on fetal growth, placental function, early postnatal growth or survivability.[46] This implies that *Nesp*, which is expressed in discrete regions of the adult brain, directly influences ongoing adult brain function, or elicits effects on adult brain function via subtle effects on neurodevelopment. However, this is not totally conclusive (for instance as discussed previously subtle effects on mother-pup interactions may give rise to large changes in the offspring when adult) and additional behavioral and neurobiological studies of this mouse, plus future work using a conditional *Nesp* knock-out, will address this issue more definitely.

What Adult Behaviors Will Imprinted Genes Influence?

One route through which imprinted genes could have evolved to influence adult brain function directly is where there is sex-biased dispersal from a social group and from this we can make a prediction that imprinted genes will influence behavior that impacts on social interactions. Asymmetries of relatedness will occur with either male or female biased dispersal,[57,58] but the most common situation in mammals is for males to disperse upon reaching sexual maturity,[62] producing "matrilineal" social groupings. If we take these as an example, given the greater sharing of maternal alleles between group members, we may expect these to promote social cohesion within the group. However, paternal alleles, which are less widely shared, will seek to limit any behavior that may reduce their presence in the next generation.

The range of behaviors that can be thought to be important to a social group is wide, taking in alarm calling, resource foraging and gathering, shared care of young and specific social cohesion behaviors such as grooming.[58,59] For instance, as *Nesp* knockout mice show a reduced propensity to explore a novel environment, we could tentatively suggest that the function of this maternally expressed gene is to promote foraging behavior. One putative neural system that may also be influenced by imprinted genes is that involved in communal care of offspring,[63] such as the oxytocin system.[64] Indeed there is some evidence that this is the case, in that *Peg3* appears to be involved in development of oxytocin neurons and maternal behavior.[26] More generally, imprinted genes may impinge on those systems involved in general affiliative behaviors, such as vasopressin.[65] In humans, it is possible that brain processes and behaviors associated with higher level social communication (i.e., speech, reading and language) may be subject to imprinted gene dependent parent-of-origin effects.[66] (For further discussion, see the chapter by Goos and Ragsdale.) By stratifying linkage data from disorders of sociality/language such as autism and dyslexia according to parental origin, we should eventually be able to identify and characterise imprinted genes affecting these important psychological constructs.

Acknowledgements

This work was supported by the Biotechnology and Biological Sciences Research Council (BBSRC), U.K. and a Babraham Institute Synergy Initiative (BBSRC, U.K.). AI is supported by the Beebe Trust and the Health Foundation. LSW is a member of the Medical Research Council (U.K.) Cooperative on Imprinting in Health and Disease.

References

1. Reik W, Walter J. Genomic imprinting: parental influence on the genome. Nat Rev Genet 2001; 2(1):21-32.
2. Tycko B. Imprinted genes in placental growth and obstetric disorders. Cytogenet Genome Res 2006; 113(1-4):271-278.
3. Davies W, Isles AR, Wilkinson LS. Imprinted gene expression in the brain. Neurosci Biobehav Rev 2005; 29(3):421-430.
4. Bassett SS, Avramopoulos D, Fallin D. Evidence for parent of origin effect in late-onset Alzheimer disease. Am J Med Genet 2002; 114(6):679-686.
5. Davies W, Isles AR, Wilkinson LS. Imprinted genes and mental dysfunction. Ann Med 2001; 33(6):428-436.
6. Ebers GC, Sadovnick AD, Dyment DA et al. Parent-of-origin effect in multiple sclerosis: observations in half-siblings. Lancet 2004; 363(9423):1773-1774.
7. Isles AR, Wilkinson LS. Imprinted genes, cognition and behaviour. Trends Cogn Sci 2000; 4(8):309-318.
8. Keverne EB, Fundele R, Narasimha M et al. Genomic imprinting and the differential roles of parental genomes in brain development. Brain Res Dev Brain Res 1996; 92(1):91-100.
9. Hager R, Johnstone RA. The genetic basis of family conflict resolution in mice. Nature 2003; 421(6922):533-535.
10. Isles AR, Baum MJ, Ma D et al. Urinary odour preferences in mice. Nature 2001; 409(6822):783-784.
11. Cassidy SB, Dykens E, Williams CA. Prader-Willi and Angelman syndromes: sister imprinted disorders. Am J Med Genet Summer 2000; 97(2):136-146.
12. Francks C, DeLisi LE, Shaw SH et al. Parent-of-origin effects on handedness and schizophrenia susceptibility on chromosome 2p12-q11. Hum Mol Genet 2003; 12(24):3225-3230.
13. Lamb JA, Barnby G, Bonora E et al. Analysis of IMGSAC autism susceptibility loci: evidence for sex limited and parent of origin specific effects. J Med Genet 2005; 42(2):132-137.
14. Yamasaki K, Joh K, Ohta T et al. Neurons but not glial cells show reciprocal imprinting of sense and antisense transcripts of Ube3a. Hum Mol Genet 2003; 12(8):837-847.
15. Yamasaki Y, Kayashima T, Soejima H et al. Neuron-specific relaxation of Igf2r imprinting is associated with neuron-specific histone modifications and lack of its antisense transcript Air. Hum Mol Genet 2005; 14(17):2511-2520.
16. Joseph R, Dou D, Tsang W. Neuronatin mRNA: alternatively spliced forms of a novel brain-specific mammalian developmental gene. Brain Res 1995; 690(1):92-98.
17. Kim J, Lu X, Stubbs L. Zim1, a maternally expressed mouse Kruppel-type zinc-finger gene located in proximal chromosome 7. Hum Mol Genet 1999; 8(5):847-854.
18. Rougeulle C, Glatt H, Lalande M. The Angelman syndrome candidate gene, UBE3A/E6-AP, is imprinted in brain. Nat Genet 1997; 17(1):14-15.
19. Albrecht U, Sutcliffe JS, Cattanach BM et al. Imprinted expression of the murine Angelman syndrome gene, Ube3a, in hippocampal and Purkinje neurons. Nat Genet 1997; 17(1):75-78.
20. Hetts SW, Rosen KM, Dikkes P et al. Expression and imprinting of the insulin-like growth factor II gene in neonatal mouse cerebellum. J Neurosci Res 1997; 50(6):958-966.
21. Davies W, Isles A, Smith R et al. Xlr3b is a new imprinted candidate for X-linked parent-of-origin effects on cognitive function in mice. Nat Genet 2005; 37(6):625-629.

22. Wang Y, Joh K, Masuko S et al. The mouse Murr1 gene is imprinted in the adult brain, presumably due to transcriptional interference by the antisense-oriented U2af1-rs1 gene. Mol Cell Biol 2004; 24(1):270-279.
23. Lee S, Walker CL, Karten B et al. Essential role for the Prader-Willi syndrome protein necdin in axonal outgrowth. Hum Mol Genet 2005; 14(5):627-637.
24. Muscatelli F, Abrous DN, Massacrier A et al. Disruption of the mouse Necdin gene results in hypothalamic and behavioral alterations reminiscent of the human Prader-Willi syndrome. Hum Mol Genet 2000; 9(20):3101-3110.
25. Kuwako K, Hosokawa A, Nishimura I et al. Disruption of the paternal necdin gene diminishes TrkA signaling for sensory neuron survival. J Neurosci 2005; 25(30):7090-7099.
26. Li L, Keverne EB, Aparicio SA et al. Regulation of maternal behavior and offspring growth by paternally expressed Peg3. Science 1999; 284(5412):330-333.
27. Jiang YH, Armstrong D, Albrecht U et al. Mutation of the Angelman ubiquitin ligase in mice causes increased cytoplasmic p53 and deficits of contextual learning and long-term potentiation. Neuron 1998; 21(4):799-811.
28. Miura K, Kishino T, Li E et al. Neurobehavioral and electroencephalographic abnormalities in Ube3a maternal-deficient mice. Neurobiol Dis 2002; 9(2):149-159.
29. Kikyo N, Williamson CM, John RM et al. Genetic and functional analysis of neuronatin in mice with maternal or paternal duplication of distal Chr 2. Dev Biol 1997; 190(1):66-77.
30. Yang H, Mattingly RR. The Ras-GRF1 exchange factor coordinates activation of H-Ras and Rac1 to control neuronal morphology. Mol Biol Cell 2006; 17(5):2177-2189.
31. Fischer-Colbrie R, Eder S, Lovisetti-Scamihorn P et al. Neuroendocrine secretory protein 55: a novel marker for the constitutive secretory pathway. Ann N Y Acad Sci 2002; 971:317-322.
32. Hinton EC, Holland AJ, Gellatly MS et al. Neural representations of hunger and satiety in Prader-Willi syndrome. Int J Obes (Lond) 2006; 30(2):313-321.
33. Boer H, Holland A, Whittington J et al. Psychotic illness in people with Prader Willi syndrome due to chromosome 15 maternal uniparental disomy. Lancet 2002; 359(9301):135-136.
34. Cassidy SB. Prader-Willi syndrome. J Med Genet 1997; 34(11):917-923.
35. Summers JA, Feldman MA. Distinctive pattern of behavioral functioning in Angelman syndrome. Am J Ment Retard 1999; 104(4):376-384.
36. Skuse DH, James RS, Bishop DV et al. Evidence from Turner's syndrome of an imprinted X-linked locus affecting cognitive function. Nature 1997; 387(6634):705-708.
37. Bishop DV, Canning E, Elgar K et al. Distinctive patterns of memory function in subgroups of females with Turner syndrome: evidence for imprinted loci on the X-chromosome affecting neurodevelopment. Neuropsychologia 2000; 38(5):712-721.
38. Davies W, Isles AR, Burgoyne PS et al. X-linked imprinting: effects on brain and behaviour. Bioessays 2006; 28(1):35-44.
39. Skuse DH. Imprinting, the X-chromosome and the male brain: explaining sex differences in the liability to autism. Pediatr Res 2000; 47(1):9-16.
40. Raefski AS, O'Neill MJ. Identification of a cluster of X-linked imprinted genes in mice. Nat Genet 2005; 37(6):620-624.
41. Cattanach BM, Kirk M. Differential activity of maternally and paternally derived chromosome regions in mice. Nature 1985; 315(6019):496-498.
42. Plagge A, Gordon E, Dean W et al. The imprinted signaling protein XL alpha s is required for postnatal adaptation to feeding. Nat Genet 2004; 36(8):818-826.
43. Lefebvre L, Viville S, Barton SC et al. Abnormal maternal behaviour and growth retardation associated with loss of the imprinted gene Mest. Nat Genet 1998; 20(2):163-169.
44. Brambilla R, Gnesutta N, Minichiello L et al. A role for the Ras signalling pathway in synaptic transmission and long-term memory. Nature 1997; 390(6657):281-286.
45. Li S, Tian X, Hartley DM et al. Distinct roles for Ras-guanine nucleotide-releasing factor 1 (Ras-GRF1) and Ras-GRF2 in the induction of long-term potentiation and long-term depression. J Neurosci 2006; 26(6):1721-1729.
46. Plagge A, Isles AR, Gordon E et al. Imprinted Nesp55 influences behavioral reactivity to novel environments. Mol Cell Biol 2005; 25(8):3019-3026.
47. Luedi PP, Hartemink AJ, Jirtle RL. Genome-wide prediction of imprinted murine genes. Genome Res 2005; 15(6):875-884.
48. Angiolini E, Fowden A, Coan P et al. Regulation of placental efficiency for nutrient transport by imprinted genes. Placenta 2006; 27 Suppl A:S98-102.
49. Barker DJ. The developmental origins of chronic adult disease. Acta Paediatr Suppl 2004; 93(446):26-33.

50. Geva R, Eshel R, Leitner Y et al. Neuropsychological outcome of children with intrauterine growth restriction: a 9-year prospective study. Pediatrics 2006; 118(1):91-100.
51. Ozanne SE, Fernandez-Twinn D, Hales CN. Fetal growth and adult diseases. Semin Perinatol 2004; 28(1):81-87.
52. Clapcott SJ, Peters J, Orban PC et al. Two ENU-induced mutations in Rasgrf1 and early mouse growth retardation. Mamm Genome 2003; 14(8):495-505.
53. Matthews K, Robbins TW. Early experience as a determinant of adult behavioural responses to reward: the effects of repeated maternal separation in the rat. Neurosci Biobehav Rev 2003; 27(1-2):45-55.
54. Bredy TW, Grant RJ, Champagne DL et al. Maternal care influences neuronal survival in the hippocampus of the rat. Eur J Neurosci 2003; 18(10):2903-2909.
55. Champagne FA, Weaver IC, Diorio J et al. Maternal care associated with methylation of the estrogen receptor-alpha1b promoter and estrogen receptor-alpha expression in the medial preoptic area of female offspring. Endocrinology 2006; 147(6):2909-2915.
56. Weaver IC, Cervoni N, Champagne FA et al. Epigenetic programming by maternal behavior. Nat Neurosci 2004; 7(8):847-854.
57. Haig D. Genomic imprinting, sex-biased dispersal and social behavior. Ann N Y Acad Sci 2000; 907:149-163.
58. Isles AR, Davies W, Wilkinson LS. Genomic imprinting and the social brain. Phil Trans Roy Soc Series B, 2006 (in press).
59. Trivers R, Burt A. Kinship and genomic imprinting. Results Probl Cell Differ 1999; 25:1-21.
60. Wilkins JF, Haig D. What good is genomic imprinting: the function of parent-specific gene expression. Nat Rev Genet 2003; 4(5):359-368.
61. Orr HA. Somatic mutation favors the evolution of diploidy. Genetics 1995; 139(3):1441-1447.
62. Chepko-Sade DB, Halpin M, Zuleyman T. Mammalian Dispersal Patterns: The Effects of Social Structure on Population Genetics: University of Chicago Press; 1987.
63. Roulin A, Hager R. Indiscriminate nursing in communal breeders: a role for genomic imprinting. Ecology 2003; 6(3):165-166.
64. Poindron P. Mechanisms of activation of maternal behaviour in mammals. Reprod Nutr Dev 2005; 45(3):341-351.
65. Curley JP, Keverne EB. Genes, brains and mammalian social bonds. Trends Ecol Evol 2005; 20(10):561-567.
66. Stein CM, Millard C, Kluge A et al. Speech Sound Disorder Influenced by a Locus in 15q14 Region. Behav Genet 2006.

CHAPTER 6

Genomic Imprinting and Human Psychology:
Cognition, Behavior and Pathology

Lisa M. Goos* and Gillian Ragsdale

Abstract

Imprinted genes expressed in the brain are numerous and it has become clear that they play an important role in nervous system development and function. The significant influence of genomic imprinting during development sets the stage for structural and physiological variations affecting psychological function and behaviour, as well as other physiological systems mediating health and well-being. However, our understanding of the role of imprinted genes in behaviour lags far behind our understanding of their roles in perinatal growth and development. Knowledge of genomic imprinting remains limited among behavioral scientists and clinicians and research regarding the influence of imprinted genes on normal cognitive processes and the most common forms of neuropathology has been limited to date. In this chapter, we will explore how knowledge of genomic imprinting can be used to inform our study of normal human cognitive and behavioral processes as well as their disruption. Behavioural analyses of rare imprinted disorders, such as Prader-Willi and Angelman syndromes, provide insight regarding the phenotypic impact of imprinted genes in the brain, and can be used to guide the study of normal behaviour as well as more common but etiologically complex disorders such as ADHD and autism. Furthermore, hypotheses regarding the evolutionary development of imprinted genes can be used to derive predictions about their role in normal behavioural variation, such as that observed in food-related and social interactions.

Genomic Imprinting in Human Cognition and Behavior

Molecular genetic studies and mouse models have clearly indicated that imprinted genes, differentially expressed from their maternally and paternally derived alleles, play a primary role in development, including nervous system development, where they influence structure, physiology and metabolism.[1-5] In addition to their developmental role, imprinted genes influence behavior, emotion and cognition across the lifespan, via functional variations in the structure and/or function of the underlying neural substrate. Their influence on behavior and cognition in humans has been demonstrated mostly through study of rare genetic syndromes such as Prader-Willi syndrome (PWS), Angelman syndrome (AS) and Russell-Silver syndrome,[6-8] although a number of more common human neuropsychiatric disorders have also shown phenotypic variations consistent with genomic imprinting effects, including Alzheimer's disease,[9] autism,[10] epilepsy,[11] schizophrenia and Huntington's disease.[12]

As knowledge of genomic imprinting has spread within the scientific community, evaluations of parental transmission patterns indicative of genomic imprinting have become relatively common

*Corresponding Author: Lisa M. Goos—Department of Psychiatry Research, The Hospital for Sick Children, 4th Floor, Black Wing, 555 University Avenue, Toronto, Ontario, Canada M5G 1X8. Email: lisa.goos@sickkids.ca

Genomic Imprinting, edited by Jon F. Wilkins. ©2008 Landes Bioscience and Springer Science+Business Media.

in familial studies as well as molecular analyses. However, knowledge of genomic imprinting remains limited among behavioral scientists and clinicians, and research regarding the influence of imprinted genes on normal cognitive processes and the most common forms of neuropathology has lagged behind that in more specialized domains.

In this chapter, we will explore how genomic imprinting may be incorporated into the study of a wide variety of normal cognitive and behavioral processes as well as their disruption. The potential influence of genomic imprinting on two very important realms of behavior, feeding and social behavior, will be explored and we will describe how knowledge of genomic imprinting effects in the brain can be used to inform our study of human cognition and psychopathology.

Genomic Imprinting and Normal Cognition

The production of genetic chimeras, or mixtures, in mice has so far been the most effective laboratory method to study the influence of imprinted genes on brain structural development (see the chapters by Davies et al and Frontera et al). By determining the regional distribution of uniparental cells within the brain, these studies have provided clues about the potential impact of imprinted genes on behavior and cognition. Chimeras are produced by inserting cells carrying only paternally derived genes (androgenetic (Ag) cells) or only maternally derived genes (parthenogenetic (Pg) or gynogenetic (Gg) cells) into a normal mouse blastocyst.[13] As long as the contribution of uniparental cells is less than 50%, the embryo is likely to survive and the developmental trajectory of the cells with the uniparental genome can be traced at dissection via cellular or genetic markers.[1,2]

Interestingly, uniparental cells are found in distinct anatomical locations in the body and the brain.[1,14] In the brain, Ag cells make their largest contribution to the mediobasal forebrain, the hypothalamus, septal nuclei and connecting neural pathways including the stria terminalis. These areas maintain energy balance and homeostasis, as well as mediating complex drive-related behaviors such as food seeking, mating, emotional expression, social aggression, circadian rhythms and the biological clock.[15,16] The hypothalamus secretes two important hormones itself, oxytocin and vasopressin, which are involved in aggression, affiliation, stress, mothering and emotion, in addition to their roles in physiological homeostasis.[17-19] Through a variety of releasing hormones, the hypothalamus also exerts control over the release of all the major endocrine hormones of the pituitary gland. Therefore, the parts of the brain to which the paternal genome makes a substantial contribution exert influence over primary motivated behaviors as well as mechanisms involved in growth and metabolism.

In contrast to Ag cells, Pg or Gg cells are virtually undetectable in the hypothalamus of mouse chimeras. Most Pg cells are found in the striatum, hippocampus and neocortex, with increasing concentrations from the occipital area to the frontal lobes. These areas underlie intelligence and most cognitive functions as well as complex emotional responses, planning and problem solving. This is necessarily a brief review of the cellular deposition patterns observed in chimeric mice. The interested reader is referred to the chapters in this volume by Davies et al and Frontera et al and to the original research papers by Keverne and colleagues.[1,20,21]

Based on the cell deposition patterns observed in chimeric mice, Goos and Silverman[22] recently investigated parent-of-origin effects in the inheritance of human cognitive abilities. Families comprised of an adult son and daughter and their biological parents were administered a neuropsychological test battery made up of tasks dependent on activity in distinct cortical areas, with multiple tasks for each area. Designations regarding localization of function were based on brain imaging and lesion studies and the aggregation of tasks for each cortical area confirmed by factor analysis. Aggregate scores for each cortical lobe were calculated and the association between the children's scores and each parent analyzed for parent-of-origin effects. Children's abilities more closely resembled the abilities of their mothers than of their fathers across the frontal, parietal and temporal lobes, with the highest correlation for the frontal lobe score. Regression analyses showed that mothers' scores were the sole significant predictor of children's score in each of these domains, with both parents' scores equally predictive for occipital lobe score.

These findings are in keeping with the Pg/Gg cell deposition pattern in chimeric mice and support the conclusion that maternal genes more strongly influence the development of the cortex in humans. There may exist one or more genes that are expressed in these brain regions for which evolution has favored a higher level of maternal expression, leading to paternal silencing. It is also possible that genes which became imprinted for reasons unrelated to intellectual function, such as a prenatal growth effect, happen to be expressed in these brain areas as well. Regardless of the original impetus, once paternally silenced, these genes may have acquired new functionality, or modifications of functionality, in the service of matrilineal inclusive fitness, an inference that is also in keeping with other data from the behavioral genetics literature.

It has been suggested that a subset of genes produce a core intellectual foundation which underlies more specialized cognitive skills.[23,24] This core accounts for about 40% of the variance between individuals on specific cognitive abilities.[25] Specialized cognitive abilities develop on top of this foundation, partly in response to unique genetic and environmental selection factors. Thus, any collection of cognitive skills has a common genetic foundation as well as developmental contributors unique to each skill and individual. If this model is correct, it may explain why numerous past attempts to measure the heritability of specific cognitive skills have shown no consistent pattern of parent-offspring correlations.[26-31] In the case of individual cognitive skills, attempts to measure heritability would be hampered by the relatively large proportion of individual variation in the skill determined by stochastic mechanisms.

The detection of genomic imprinting effects via familial resemblances was mathematically modeled by Spencer,[32] who predicted that high standard errors would obscure any findings of interest if the method were applied to the study of phenotypic traits. Pooling the data from a number of different cognitive tasks, the method used by Goos and Silverman, effectively strengthens the proportion of variation explained by common underlying mechanisms relative to the random variation associated with each individual task. Thus, the influence of the common underlying foundation becomes more readily discernable. Taken together, the evidence suggests that this intellectual foundation may be particularly sensitive to the influence of one or more maternally expressed imprinted genes.

Of course, there are other possible explanations for why specific abilities in children might preferentially resemble those of their mothers. For instance, mothers spend more time with children than do fathers in traditional families and may be presumed, thereby, to take a greater role in their cognitive development. While there is no basis to theorize that this would produce the pattern of parental differences predicted and obtained in the Goos and Silverman study, a replication including nonbiologically related offspring would be instructive, as would studies of functions presumed to be regulated by the portions of the brain to which the paternal contribution is primary, such as homeostatic mechanisms, reproductive and primary motivated behaviors. Certainly, further studies of parent-of-origin effects in cognition are warranted.

Genomic Imprinting and Social Skills

Complex social behavior, including language, is often cited as the most defining characteristic of the modern human species.[33] Humans understand, respond to and manipulate the behavior of conspecifics more than any other animal. Two highly correlated skill sets enable this behavior:[34] executive control abilities, (including working memory, behavioral inhibition and cognitive flexibility) which support planned, purposeful behavior;[35] and 'theory of mind,' which underlies the ability to understand the thoughts, feelings and intentions of others.[36] Theory of mind is also important in using and understanding certain aspects of language[37] and often includes the concept of empathy (for a comprehensive discussion of how the term 'empathy' is used see Preston and de Waal, 2002[38]). All of these skills are associated with cortical activity.[39-41]

It has been hypothesized that the evolution of the human brain, especially the enlarged neocortex, was driven by the need to manage increasingly large social networks.[42] Whereas executive control contributes to all higher cognitive functions to some degree, both social and nonsocial, empathy and theory of mind are specifically associated with social competence, making them excellent phenotypic candidates in the study of genetic influences on human brain evolution and development.

The strongest evidence suggesting that imprinted genes affect human social abilities comes from research on autistic spectrum disorder (AD) and Turner's syndrome (TS). AD is associated with impairment of both executive function and theory of mind, and is characterized by both social and nonsocial deficits with or without accompanying mental retardation. A related syndrome—Asperger's—is sometimes referred to as 'high functioning autism' due to the presence of impaired social functioning without the mental retardation or language impairment typically seen in AD. The social deficits in both include inability to understand social situations and difficulty interpreting facial expressions. The nonsocial deficit in AD is typified by highly repetitive, inflexible behavior.

AD is known to be highly heritable[43] and large scale genetic linkage studies have found several loci that are potentially associated with AD,[10,44-55] some of which are also associated with parent-of-origin effects.[56] More specifically, AD has been associated with maternally expressed loci on chromosome 15,[57-61] paternally expressed loci on chromosome 7[10,62] and the X chromosome.[63] Although no phenotype-genotype correlations have been obtained as yet, there is also evidence that linguistic, social and nonsocial components of the AD phenotype are independently influenced by different genes.[64,65]

Turner's syndrome occurs when one of the two X chromosomes normally found in females is missing or contains structural defects. Depending on the origin of the functional X chromosome, TS females express either the maternally or paternally derived X-linked genes, designated X_m or X_p respectively. Poor social skills are common in TS girls and they are at least 200 times more likely than otherwise normal individuals to have AD.[66] Both individuals with AD and those with TS fail to recognize facial expressions of fear or ascertain gaze direction, both components of theory of mind.[67,68] The similarities in social skill deficits seen in AD and TS suggest there are common biological underpinnings in the two disorders. Studies of parent-of-origin effects in TS indicate the commonalities may be due to the influence of imprinted, X-linked gene(s).

Structural and functional comparisons of girls with X_m, girls with X_p and normal controls suggest that parental origin of the X chromosome is an important factor mediating variability among these groups. For example, brain imaging studies have found evidence that X_m TS females have increased superior temporal gyrus volumes (putatively involved in interpreting emotion from eye contact[69]) compared to X_p TS females and controls.[70] This ability is thought to be fundamental to theory of mind and social competence.[71] Skuse and colleagues[72] found that X_p TS girls had better social skills, verbal skills, planning ability and behavioral inhibition than X_m TS girls and normal girls had better social skills and behavioral inhibition than normal boys. Based on these observations in TS, they proposed a 'threshold liability model of autism' in which genes expressed specifically from the X_p enhance social cognition and protect daughters from developing AD.[73] Since sons cannot normally inherit X_p, this model proposes to account not only for the biased sex ratio in AD (4 boys: 1 girl), but also the extreme bias observed in Asperger's syndrome (10 boys: 1 girl), in which social impairment is the primary (and perhaps only) deficit.

Overall, the results from the studies of AD and TS suggest that there are one or more X-linked, imprinted genes influencing theory of mind and social skill. Multiple imprinted X-linked genes involved in the etiology of autism, uniquely influencing social and nonsocial traits, is not as unlikely as it might first appear, as there is evidence of a concentration of genes affecting cognition on the X chromosome.[74] Supporting evidence also comes from mouse studies: Raefski and O'Neill[75] have discovered a cluster of X-linked imprinted genes in mice and Davies and colleagues[76] have found an X-linked, imprinted gene affecting cognitive function. There are also likely to be autosomal imprinted genes influencing social behavior, since there are autosomal imprinted genes associated with AD.

Evolutionary Interpretations

According to the kinship theory of imprinting, preferential selection of maternally expressed genes, particularly those governing resource transfer prenatally, would have functioned to protect maternal reproductive fitness from excess resource demands generated by mechanisms under paternal genetic control.[77] However, the extended postnatal development period and sociality characteristic of humans suggests that opportunities for conflict and fitness effects after birth will extend beyond the parents to matrilineal and patrilineal kin.[78] Accordingly, imprinting of the genes underlying a number of postnatal behaviors may be expected.

In chimeric mice, the hypothalamus and cortex show differential expression of parental genes and both structures are involved in the instigation and execution of motivated behavior. Thus, parental conflict over the performance of motivated behavior may have given rise to maternal gene expression in the cortex to counter paternal expression in the hypothalamus and related structures. Substantial and distinctive neural and behavioral interconnectivity between these two structures supports this supposition.[79,80]

Given the social nature of primates and the importance of the social group in survival and reproductive success,[16,81] this conflict would have been waged within a social context, exerting positive selection on the genes underlying social behavior. As a result, social skills are likely to be influenced by both maternally and paternally imprinted genes.

It has already been suggested by primate anthropologists that increasing social pressure drove the development of the neocortex, particularly the frontal cortex, which subserves skills such as the ability to form complex associations, make transitive inferences and predict the behavior of fellow group members.[82,83] Studies of brain size across species suggest that the consistent increase in neocortex size relative to group size observed in primates[84] mostly reflects increasing social skills as opposed to visual recognition skills,[85] home-range size, tool use,[86] diet, number of males in the group or body size.[14]

It may be plausibly speculated that once a significant number of genes were maternally expressed in the cortex, these genes were subject to further evolutionary pressures based on new social and environmental challenges.[82,87] Thus, emergent nonsocial cognitive skills would have retained their maternal genetic basis, resulting in the mother's continued influence on cognitive development in offspring, as observed by Goos and Silverman.

If increasingly complex sociality drove the evolution of maternally and paternally imprinted genes in the mammalian brain, any disruption of the balance between the maternally influenced cortex and the paternally influenced limbic system could lead to deficits in a range of social skills, a hypothesis comprehensively explored by Badcock and Crespi.[56] This mechanism may also help explain some of the etiological complexity in AD, including evidence of cortical or neuroendocrine dysfunction, but not necessarily both.[88-94] An imbalance may be due to mutations in the genes or the imprinting process, or it may simply reflect natural variation and/or polymorphism in the influence of maternally and paternally derived genes on social behavior.

There is mounting evidence that AD is not a discrete pathology, but the extreme end of a continuum that encompasses the entire normal population. Baron-Cohen[95] has characterized the social and nonsocial deficits in AD as extreme forms of two distinct, measurable traits that are normally distributed in the population:[64] empathizing, which includes theory of mind and systemizing. In this model, individuals with AD form an extreme cluster in the normal distribution, with extremely low empathizing and extremely high systemizing scores. Evidence that the social, linguistic and nonsocial components of the AD phenotype may be influenced by different genes, with non-additive effects (which could include imprinting) being more influential at the extremes supports this view, as does the fact that the various components of the AD phenotype have been found to 'splinter' and occur separately in close relatives of the affected individual.[64,65] The alternative explanation—that these families possess several imprinting errors that come together in the member diagnosed with AD—is unlikely given the relative frequency of AD in the population.

If AD is the result of normal allelic variation, one may ask how such deleterious alleles remain in the population. In this context it is worth distinguishing the meaning of 'pathology' within psychiatry as opposed to evolutionary biology. In psychiatry, a common working usage is to view behavior as pathological when it becomes very difficult for a person to cope with everyday life.[96] This somewhat subjective and culture-bound definition contrasts with the simple view from the evolutionary perspective: it's all good as long as you are reproductively fit. Furthermore, a phenotype associated with reduced individual reproductive fitness may still persist in a population if it is also associated with increased fitness in close kin.

It has been suggested that the nonsocial component of the AD phenotype, systemizing, is associated with proficiency in areas such as engineering and computer science.[95] This trait perhaps had its equivalent during early human evolution in, for example, stone-tool use. 'Mild' AD and Asperger's syndrome might themselves have (or have had) fitness benefits to the individual or their kin due to strong systemizing skills. It could be further speculated that the associated impairment in social skills enables more focused application of systemizing skills. Since systemizing skills are increasingly useful in the modern workplace, it has been proposed that assortative mating between high systemizers is producing increasing numbers of offspring with AD.[97]

Any trait that is influenced by imprinted genes may have been selected via pressures resulting from tension between maternal (or matrilineal) and paternal (or patrilineal) interests. Knowledge of such influence begs a re-appraisal of complex traits for which familial patterns cannot be explained by classical Mendelian genetics. For example, a comprehensive analysis of the data on personality from the Minnesota Twin Family Registry[98] found a non-additive (i.e., non-Mendelian) genetic component to be significant in 10 of 14 measures. The environmental and social pressures which lead to differences in maternal and paternal interests may have changed, in some cases dramatically, over our evolutionary history and especially in relatively recent modern human history. Behaviors which once conferred fitness may no longer do so.

Imprinted Syndromes, Behavioral Phenotypes and Neuropsychological Research

The recognition of a distinctive behavioral phenotype associated with a syndrome or condition of known genetic etiology has the potential to guide us towards genes that contribute to that behavior in the general population.[99,100] Many imprinted syndromes, including AS and PWS, have distinctive behavioral profiles. PWS and AS are commonly presented models of the influence of genomic imprinting on development and behavior, as both are caused by the loss of gene expression on chromosome 15 at 15q11-q13.[7] Distinct phenotypic differences between the syndromes are determined by the parental origin of the genetic mutation: AS is caused by the loss of maternally expressed genes in this region, whereas the loss of paternally expressed genes is implicated in PWS. The consequences of PWS and AS, particularly on postnatal metabolism, are also discussed in the chapter in this volume by Frontera et al.

PWS is characterized by mild to moderate mental retardation, hypotonia (lack of muscle tone and response to stretch), hypothalamic hypogonadism, growth hormone deficiency, poor temperature regulation and obesity.[99,101] Infants with PWS show prenatal growth retardation, poor suck reflexes following birth and often show failure to thrive as a result.[102] In the older child, food-related behavioral problems continue to occur, specifically insatiable appetite, food stealing, gorging and the consumption of nonfood items.[103] The hyperphagia may be very extreme, with weight gains of more than 200% above normal body weight.[7,101]

The behavioral phenotype characteristic of PWS has certainly helped us learn more about the genetic and physiological underpinnings of feeding, appetite and satiety. Other aspects of the PWS phenotype may do the same for behavioral and clinical scientists. Using PWS as a model, the following discussion will demonstrate how phenotypic traits identified in imprinted syndromes can inform the study of other, relatively more common forms of behavior or neuropsychiatric dysfunction.

Genomic Imprinting, Feeding Behavior and Health

Clearly behavior plays an important role in obesity and the health problems it engenders; it is not possible to explain the recent and rapidly increasing secular trend in obesity by genetics alone. While it may be tempting to consider the causes of obesity to be gluttony or sloth in various combinations of degree, strong familial patterns of obesity imply genetic components as well. In fact, over 250 quantitative trait loci have been associated with obesity, implying multiple opportunities for gene-environment interactions.[104] Linkage studies of obesity show some evidence of parent-of-origin effects, both maternal and paternal.[105,106]

Studies of PWS have helped us to learn more about the role of the hypothalamus, neurotransmitters and imprinted genes in the regulation of feeding behavior. Lesions of the paraventricular nucleus of the hypothalamus in mice lead to a similar behavioral obesity syndrome, in which they overeat to the point of obesity despite the lack of an associated metabolic change.[107] Mice with a mutation of the serotonin 2C receptor subtype gene show a similar propensity,[108] apparently due to the loss of inhibitory control over neuropeptide Y, a potent stimulator of hunger and food intake.[109]

Haig and Wharton[110] have interpreted the feeding behavior of PWS children as reflecting maternal resource-preserving interests taken to the extreme. Carrying only maternally inherited genes in the PWS region, these children are undemanding in utero and on breast milk, but are voracious foragers and consumers of food once weaned. It may be reasonable to infer the normal function of the gene(s) lacking expression in PWS by inverting the PWS phenotype. This suggests that the normal activity of the paternally inherited alleles missing in PWS serve to promote fetal growth and postnatal suckling, but lead to a moderate and discriminating appetite in weaned infants and children. As Haig and Wharton note, the indiscriminate foraging seen in PWS, where quantity is the overarching criterion of food choice, contrasts startlingly with the everyday experience of many exasperated parents, chasing the pursed lips of their infants with spoons of wholesome food before finally producing the longed-for yogurt or jelly.

In addition to their conflict over the perinatal demand for maternal resources, maternally and paternally inherited alleles may have divergent interests in food preferences. There has been speculation that the high proportions of fat and sugar in popular junk food reflect food preferences that would have motivated stone-age humans to seek out these energy rich resources at a time when they were hard to come by. However, patrilineal interests might preferentially be served by eating sugar. After eating a high-sugar food, blood sugar rises for a short time then plummets as insulin is released, shutting down the body's production of glucose. When blood sugar is scarce, priority goes to the limbic system rather than the cortex, thus favoring the paternally influenced part of the brain.

Insulin directly affects appetite and feeding via insulin receptors in the brain[111] and there is some evidence that the insulin gene (*INS*) is imprinted, although studies do not agree.[112-114] In some populations, diabetes shows a parent-of-origin effect[115] and it is possible that *INS* expression is polymorphic. The insulin gene is paternally expressed in human and murine yolk sac[116] and insulin has a role in embryogenesis (as yet unclear) distinct from its effect on adult carbohydrate metabolism.[117]

Related imprinted genes are known to influence fetal growth and energy balance, namely the first imprinted genes to be discovered: insulin-like growth factor II (*Igf2*),[118] its receptor (*Igf2r*)[119] and *H19*, a regulatory element.[120] In humans, *IGF2* and *H19* expression is usually complementary, being maternally and paternally silenced, respectively, but they are dissociated and follow different patterns in fetal and adult brain tissues.[3] It is unclear whether imprinted genes regulating growth also influence feeding-related behaviors, but changes in imprinting status specifically expressed in the brain suggest this may be the case. Indeed, *IGF2* has been considered a candidate gene in eating disorders.[121] It remains to be seen whether any other major players in appetite control or feeding behavior are imprinted. Leptin, for example, is very important for long-term appetite control[122] and its gene maps to 7q31.1,[123] which is flanked by imprinted regions. It is possible that growth metabolism and growth-oriented behaviors (feeding, satiety, nursing, appetite) are linked via the same imprinted genes.

Perhaps most interestingly, the relationship between imprinting and feeding behavior is not a one-way street. Research in mice has provided evidence that maternal diet can alter gene methylation status, imprinting and associated adult-onset phenotypes. The *agouti* gene, which produces a yellow coat color, is silenced in offspring whose mothers are fed a methyl-donor rich diet pre- and postnatally. Apart from their tell-tale brown coat, offspring also have reduced susceptibility to obesity, diabetes and cancer.[124] Human mothers have long been advised to take folic acid and B12 supplements (both methyl-donors) to ensure a healthy pregnancy. Further studies suggest that the diet of offspring may also influence susceptibility to obesity and diabetes. Weaned mouse pups fed a methyl-donor deficient diet were subject to a loss of imprinting of the *Igf2* gene.[125] Other differentially methylated regions were not affected, suggesting that diet does not affect imprinting in general, but is specific to genes involved in growth and metabolism. Furthermore, this effect was not reversed during a subsequent 100 days on a normal diet.

Similarly, Gardner and Lane[126] found that female mice fed a high protein diet produced embryos with abnormal imprinting. Some of the genes involved in early embryo development were inactivated and implantation was also affected. Alarming news soundbites warning women on the increasingly popular Atkins diet that they were courting infertility were soon issued in response. Despite the research cited above, it is worth bearing in mind the considerable differences between the natural diets of mice and humans before issuing dire warnings of this kind. One of the popular marketing ploys used to promote the high protein, low carbohydrate diet is to claim that it reflects the diet natural to our hunting and gathering, pre-agricultural ancestors. There is probably some truth in this and although it would be highly surprising to find it rendered our distant forebears infertile, there may well be other repercussions to which today's longer lived and relatively sedentary humans are more susceptible.

In humans, it is known that low birth weight is a risk factor for adult obesity and diabetes. Examination of records from famines such as the Dutch Famine (1944-1945) show that poor nutrition in early gestation leads to obesity in later life.[127] This effect is established early in fetal development and mediated by long-term metabolic regulation by the hypothalamus, producing a 'thrifty phenotype.'[128] This phenotype is characterized by restricted brain growth and metabolic programming suited to poor nutrition, for example, by storing any excess nutrition as fat. When food supply is plentiful, such a strategy may prove unnecessary and lead to obesity. As in the mouse studies cited above, this type of fetal programming may be regulated via epigenetic mechanisms including imprinting.[129] It is not known if or how these factors influence food preferences, as seen in PWS. Future studies might look at the relationship between birth size, early neonatal growth and postweaning food preferences. Since this is a profile of normal behavioral variation (unlike the case of PWS) it would be appropriate to collect the same data for first-degree relatives.

Human feeding behavior, obesity and related metabolic conditions present somewhat of a paradox: they have strong genetic components, yet show rapid changes from one generation to the next. The evidence that not only can imprinted genes affect feeding behavior, but feeding behavior and diet can influence imprinting provides a mechanism by which environmental factors might play a role in shaping behavior and preferences as well as susceptibilities to obesity and diabetes. And while comparison with the probable diet of our prehistoric ancestors can explain much about our present-day food preferences, tension between maternal and paternal gene expression might account for some additional variation in these traits.

Genomic Imprinting and Psychopathology

OCD

In addition to abnormal feeding behavior, PWS children also commonly display a triad of behavioral problems involving obsessive and compulsive tendencies, temper issues and externalizing behavior and emotional problems.[99] These features are comorbid with, but distinct from, other aspects of the phenotype. Obsessive Compulsive Disorder (OCD) is characterized by recurring obsessions and/or compulsions and has been estimated to affect nearly 5 million people in the United States alone.[130] Evidence for a strong genetic component to OCD has emerged from twin, family and segregation studies.[131]

Studies of OCD in PWS probands have found levels of severity in keeping with those observed in children clinically diagnosed with OCD and considerably higher than that found in individuals with heterogeneous mental retardation.[99] The phenotype-genotype correlation in PWS with OCD was confirmed in a comparison of PWS and 'PWS-like' patients, who presented all the features of PWS but lacked the 15q11-q13 genotype. OCD behaviors were only observed in the PWS patients with the genetic diagnosis, despite a lack of group differences in obesity, IQ, food-related difficulties or overall maladaptive behavioral problems.[132]

To our knowledge, association studies of genes in the PWS critical area and OCD have not been conducted to date, however there are sound behavioral and physiological reasons for doing so. In addition to the behavioral overlap, both OCD and PWS have been associated with increased cerebrospinal oxytocin levels.[133,134] The success of serotonin reuptake inhibitors in the treatment of OCD also suggests serotonergic involvement and a number of serotonin system genes have been linked with OCD, including the serotonin 2A receptor gene[135] and the serotonin transporter.[136] Genes within the PWS region also influence serotonergic function, specifically the serotonin 2C receptor.

Serotonin 2C receptor subtypes (designated 5-HT2CR) are found widely distributed throughout the brain and spinal cord.[15] The *5-HT2CR* gene appears influential in the food-related aspects of the PWS phenotype,[107-109] and has been implicated in a mouse model and pharmacological studies of OCD.[137] Thus, function of the 5-HT2C receptor may contribute to both PWS and OCD. However, it is unlikely this link is due to *5-HT2CR* gene polymorphisms, as it is located on the X chromosome in humans,[138] and neither PWS nor OCD show the pattern of inheritance characteristic of sex linkage.

Cavaille and colleagues[138] have identified a unique subset of small nucleolar RNA transcripts, or snoRNAs, that link 5-HT2CR dysfunction to PWS. SnoRNAs normally function throughout the body in the modification of ribosomal RNA (rRNA) during ribosome synthesis. These particular snoRNAs, however, show no association with rRNA. Instead, they are found only in brain tissue and appear to function in the modification of the 5-HT2CR mRNA prior to synthesis of the functional receptor. The genes encoding three of these snoRNAs are paternally expressed and located within the PWS-critical region on chromosome 15.[138-140] Furthermore, the genes encoding these three snoRNAs are not expressed in brain tissue from PWS patients or in mouse models of the disease.[138] It seems reasonable to speculate that these genes may also be involved in the etiology of OCD, at least in some cases. Future behavioral and molecular analyses will be necessary to evaluate this hypothesis.

ADHD

Two additional aspects of the PWS behavioral phenotype may provide useful clues to the etiology of Attention Deficit Hyperactivity Disorder (ADHD). ADHD is one of the most common heritable mental health disorders of childhood.[141] Generally characterized by impairing levels of inattention and/or hyperactivity-impulsivity,[142,143] children with ADHD also display a range of comorbid psychopathologies including disruptive behavior disorders such as conduct disorder (CD)[144-148] and depression.[149-152]

Evidence suggests that ADHD comorbid with CD or mood disorders, particularly bipolar affective disorder (BPAD), are distinct etiological entities, with unique genetic underpinnings relative to ADHD alone.[146,153-160] A wide variety of genes are implicated in ADHD, but substantial phenotypic and genetic heterogeneity among ADHD probands has complicated the search for causative factors. Phenotypic overlap with PWS may help identify the unique genetic underpinnings associated with CD and BPAD, both in ADHD probands and in the general population.

Wigren and Hansen[161] found that 38% of PWS children also met criteria for ADHD, CD, or both and CD is one of the major problems encountered by parents of PWS children.[162] PWS is also commonly associated with recurrent or cyclical bouts of depression.[163,164] Hypothalamic dysfunction and the disruption of the paternal 15q11-q12 region have also been associated with depression in individuals without the full PWS diagnosis.[165,166]

Although imprinting effects have rarely been included in studies of ADHD, there is evidence to suggest it may play an important role. Hyperactivity in mice was the first behavioral effect of imprinted genes to be documented,[167] and a number of human neurological disorders that commonly occur in conjunction with ADHD show variations in symptoms or severity based on the parent-of-origin, including Tourette's syndrome[168] and BPAD.[169] In addition, a number of genes relevant to the functioning of the dopamine and serotonin neurotransmitter systems, which are the usual targets of pharmacological intervention in ADHD, CD and depression, have shown preferential transmission of paternal alleles: the gene for dopa decarboxylase enzyme (DDC, 7p11), which catalyzes the synthesis of dopamine;[170,171] tryptophan hydroxylase 2 ($TPH2$, 12q21.1), a brain-specific, rate-limiting enzyme in serotonin synthesis;[172-176] and brain-derived neurotrophic factor ($BDNF$, 11p13), a growth factor that interacts with serotonin to regulate growth, survival and plasticity of serotonergic neurons.[177-179]

Recently, parent-of-origin effects in the transmission of depression within ADHD probands were reported, with higher levels of self-rated depression in the children of fathers with a history of ADHD.[180] Heritable parent-of-origin effects in the transmission of CD have also been documented, with paternal antisocial personality disorder predicting offspring CD and both conditions influencing an individual's susceptibility to drug and alcohol dependence.[181-185] In an analysis of 42 genes implicated in ADHD, CD and oppositional defiant disorder, Comings and colleagues[186] found that CD was preferentially associated with hormone and neuropeptide genes, in contrast to the other disorders. Thus, hypothalamic involvement may be the pathophysiological link between CD and PWS. Given the importance of paternally expressed genes in the neurobiology of PWS and hypothalamic development in Ag chimeras, perhaps imprinted genes play a role in CD as well.

Behavioral commonalities among disorders may or may not be due to common underlying causal factors, genetic or otherwise. However, a great deal of research in ADHD is focused on the identification of more etiologically homogeneous phenotypes for analysis, in order to improve the power of molecular tests to detect relevant genes.[187-191] This strategy has met with some success using comorbidities,[192,193] cognitive abilities,[188] DSM subtype,[194-196] and proband sex[197-199] as grouping variables, among others. Further segregation of the complex ADHD phenotype on the basis of commonalities with known imprinted disorders such as PWS, or the study of parent-of-origin effects, may be another effective way to proceed.

Conclusion

Research in traditional psychiatric genetics involves searching for unknown genes in psychiatric phenotypes.[200] Using this approach, progress has been slow despite considerable advances in molecular analysis techniques and technology. In the case of most complex psychological disorders this is due to polygenic inheritance, pleiotropy and epistatic interactions, which lead to low statistical power in molecular analyses. A growing body of evidence suggests that incorporating both genetic and epigenetic views will be more successful.[201] By combining genetic and epigenetic perspectives of etiology, studies of complex disorders such as autism, ADHD and schizophrenia can gain specificity and power. To be most successful, however, the combined efforts of molecular, behavioral and clinical scientists using this approach will be required.

Due to their unusual form of inheritance, imprinted syndromes are relatively well understood genetically and physiologically. Patterns of similarity between these syndromes and more common psychological disorders may direct our attention to less obvious genetic factors with important etiological roles and contribute to our understanding of specific functional systems in typically developing persons as well as disease processes.[202] Even if the specific genes involved are as yet unknown, such an approach can significantly narrow the range of possibilities.

Our understanding of the role of imprinted genes in behavior lags far behind our understanding of their roles in perinatal growth and development. The significant influence of genomic imprinting during development sets the stage for structural and physiological variations affecting psychological function and behavior, as well as other physiological systems mediating health and wellbeing, such as addictions,[203] cancer, allergic and immune responses,[204-207] Alzheimer's disease,[9,208-210] energy metabolism and glucose tolerance[112,211-213] (see also the chapter by Frontera et al) and cardiovascular disease.[214,215] While not all of these disorders have shown evidence of genomic imprinting as yet, all are associated with genes that are imprinted during development and which may contribute to disease susceptibility.

All too often, the queries of behavioral scientists and clinicians regarding unique parental contributions to cognition or behavior are viewed suspiciously or dismissed out of hand in favor of explanations involving environmental factors or learning processes. As more and more imprinted genes that influence behavior and cognition are identified, studies of parent-of-origin effects in behavioral phenotypes will be a necessary component of a complete research program. Guided by sound theoretical underpinnings and methodology, behavioral studies of parent-of-origin effects should be included as one more tool to decipher the complex etiological factors underlying most psychological processes.

References

1. Keverne EB, Fundele R, Narasimha M et al. Genomic imprinting and the differential roles of parental genomes in brain development. Devel Brain Res 1996; 92:91-100.
2. Keverne EB. Genomic imprinting in the brain. Curr Op Neurobiol 1997; 7:463-468.
3. Pham NV, Nguyen MT, Hu JF et al. Dissociation of IGF2 and H19 imprinting in human brain. Brain Res 1998; 810(1-2):1-8.
4. Plagge A, Isles AR, Gordon E et al. Imprinted nesp55 influences behavioral reactivity to novel environments. Mol Cell Biol 2005; 25(8):3019-3026.
5. Polychronakos C, Kukuvitis A, Giannoukakis N et al. Parental imprinting effect at the INS-IGF2 diabetes susceptibility locus. Diabetologia 1995; 38(6):715-719.
6. Whittington J, Holland A, Webb T et al. Cognitive abilities and genotype in a population-based sample of people with Prader-Willi syndrome. J Intellect Disabil Res 2004; 48(Pt 2):172-187.
7. Flint J. Implications of genomic imprinting for psychiatric genetics. Psychol Med 1992; 22:5-10.
8. Davies W, Isles AR, Wilkinson LS. Imprinted genes and mental dysfunction. Ann Med 2001; 33(6):428-436.
9. Bassett SS, Avramopoulos D, Fallin D. Evidence for parent of origin effect in late-onset Alzheimer disease. Am J Med Genet 2002; 114(6):679-686.
10. Lamb JA, Barnby G, Bonora E et al. Analysis of IMGSAC autism susceptibility loci: evidence for sex limited and parent of origin specific effects. J Med Genet 2005; 42(2):132-137.
11. Ottman R, Annegers JF, Hauser WA et al. Higher risk of seizures in offspring of mothers than of fathers with epilepsy. Am J Hum Genet 1988; 43:257-264.
12. Isles AR, Wilkinson LS. Imprinted genes, cognition and behavior. Trends Cog Sci 2000; 4(8):309-318.
13. Barton SC, Ferguson-Smith AC, Fundele R et al. Influence of paternally imprinted genes on development. Development 1991; 113:679-688.
14. Keverne EB, Martel FL, Nevison CM. Primate brain evolution: genetic and functional considerations. Proc Roy Soc Lond, B 1996; 262:689-696.
15. Kandel ER, Schwartz JH, Jessell TM. Essentials of Neural Science and Behavior. Norwalk, Connecticut: Appleton and Lange, 1995.
16. Hole JW. Human Anatomy and Physiology. Dubuque, Iowa: Wm C Brown, 1984.
17. Ferris CF. Vasopressin/oxytocin and aggression. Novartis Found Symp 2005; 268:190-198.
18. Cushing BS, Kramer KM. Mechanisms underlying epigenetic effects of early social experience: the role of neuropeptides and steroids. Neurosci Biobehav Rev 2005; 29(7):1089-1105.
19. Fries AB, Ziegler TE, Kurian JR et al. Early experience in humans is associated with changes in neuropeptides critical for regulating social behavior. Proc Natl Acad Sci 2005; 102(47):17237-17240.
20. Allen ND, Logan K, Lally G et al. Distribution of parthenogenetic cells in the mouse brain and their influence on brain development and behavior. Proc Natl Acad Sci 1995; 92:10782-10786.
21. Keverne EB. Molecular genetic approaches to understanding brain development and behavior. Psychoneuroendocrinology 1994; 19:407-414.
22. Goos LM, Silverman I. The inheritance of cognitive skills. Does genomic imprinting play a role? J Neurogenet 2006; 20:19-40.
23. Flint J. The genetic basis of cognition. Brain Res Bull 1999; 122:2015-2031.
24. Plomin R, Hill L, Craig IW et al. A genome-wide scan of 1842 DNA markers for allelic associations with general cognitive ability: A five stage design using DNA pooling and extreme selected groups. Behav Genet 2001; 31(6):497-509.
25. Jensen AR. The puzzle of nongenetic variance. In: Sternberg RJ, Grigorenko EL, eds. Intelligence, Heredity and Environment. Cambridge: Cambridge University Press, 1998:42-88.
26. DeFries JC, Ashton GC, Johnson RC et al. Parent-offspring resemblance for specific cognitive abilities in two ethnic groups. Nature 1976; 261(5556):131-133.
27. DeFries JC, Johnson RC, Kuse AR et al. Familial resemblance for specific cognitive abilities. Behav Genet 1979; 9(1):23-43.
28. Loehlin JC, Sharan S, Jacoby R. In pursuit of the "spatial gene": a family study. Behav Genet 1978; 8(1):27-41.

29. McGee MG. Intrafamilial correlations and heritability estimates for spatial ability in a Minnesota sample. Behav Genet 1978; 8(1):77-80.
30. Park J, Johnson RC, DeFries JC et al. Parent-offspring resemblance for specific cognitive abilities in Korea. Behav Genet 1978; 8(1):43-52.
31. Spuhler KP, Vandenberg SG. Comparison of parent-offspring resemblance for specific cognitive abilities. Behav Genet 1980; 10(4):413-418.
32. Spencer HG. The correlation between relatives on the supposition of genomic imprinting. Genetics 2002; 161:411-417.
33. Crow TJ. Introduction. In The Speciation of Modern Homo sapiens. Proc Br Acad 2002; 106:1-20.
34. Hughes C. Executive function in preschoolers: links with theory of mind and verbal ability. Brit J Dev Psychol 1998; 16:233-253.
35. Welsh M, Pennington B, Groisser D. A normative-developmental study of executive function. Devel Neuropsych 1991; 7:131-149.
36. Premack D, Woodruff G. Does the chimpanzee have a theory of mind? Behav Brain Sci 1978; 1:515-526.
37. Dunbar RIM. On the origin of the human mind. In: Caruthers P, Chamberlain A, eds. Evolution and the Human Mind: Modularity, Language and Meta-Cognition. Cambridge: Cambridge University Press, 2000:238-253.
38. Preston SD, de Waal FB. Empathy: Its ultimate and proximate bases. Behav Brain Sci 2002; 25(1):1-20.
39. Perry RJ, Rosen HR, Kramer JH et al. Hemispheric dominance for emotions, empathy and social behavior: evidence from right and left handers with frontotemporal dementia. Neurocase 2001; 7(2):145-160.
40. Shallice T. 'Theory of mind' and the prefrontal cortex. Brain 2001; 124(Pt 2):247-248.
41. Keysers C, Wicker B, Gazzola V et al. A touching sight: SII/PV activation during the observation and experience of touch. Neuron 2004; 42(2):335-346.
42. Dunbar RIM. The social brain hypothesis. Evol Anth 1998; 6:178-190.
43. Steffenburg S, Gillberg C, Hellgren L et al. A twin study of autism in Denmark, Finland, Iceland, Norway and Sweden. J Child Psychol Psychi 1989; 30(3):405-416.
44. Auranen M, Vanhala R, Varilo T et al. A genomewide screen for autism-spectrum disorders: evidence for a major susceptibility locus on chromosome 3q25-27. Am J Hum Genet 2002; 71(4):777-790.
45. Buxbaum JD, Silverman JM, Smith CJ et al. Evidence for a susceptibility gene for autism on chromosome 2 and for genetic heterogeneity. Am J Hum Genet 2001; 68:1514-1520.
46. Collaborative Linkage Study of Autism (CLSA). An autosomal genomic screen for autism. Am J Med Genet (Neuropsychiatric Genetics) 1999; 88:609-615.
47. International Molecular Genetic Study of Autism Consortium (IMGSAC). A full genome screen for autism with evidence for linkage to a region on chromosome 7q. Hum Mol Genet 1998; 7:571-578.
48. International Molecular Genetic Study of Autism Consortium (IMGSAC). A genomewide screen for autism: strong evidence for linkage to chromosomes 2q, 7q and 16p. Am J Hum Genet 2001; 69:570-581.
49. Liu J, Nyholt DR, Magnussen P et al. A genomewide screen for autism susceptibility loci. Am J Hum Genet 2001; 69(2):327-340.
50. Philippe A, Martinez M, Guilloud-Bataille M et al. Genome-wide scan for autism susceptibility genes. Paris Autism Research International Sibpair Study. Hum Mol Genet 1999; 8(5):805-812.
51. Risch N, Spiker D, Lotspeich L et al. A genomic screen of autism: evidence for a multilocus etiology. Am J Hum Genet 1999; 65(2):493-507.
52. Yonan AL, Alarcon M, Cheng R et al. A genomewide screen of 345 families for autism-susceptibility loci. Am J Hum Genet 2003; 73(4):886-897.
53. Alarcon M, Yonan AL, Gilliam TC et al. Quantitative genome scan and Ordered-Subsets Analysis of autism endophenotypes support language QTLs. Mol Psychiatry, 2005.
54. Cantor RM, Kono N, Duvall JA et al. Replication of Autism Linkage: Fine-Mapping Peak at 17q21. Am J Hum Genet 2005; 76(6).
55. Vorstman JA, Staal WG, van Daalen E et al. Identification of novel autism candidate regions through analysis of reported cytogenetic abnormalities associated with autism. Mol Psychiatry 2006; 11(1):1, 18-28.
56. Badcock C, Crespi B. Imbalanced genomic imprinting in brain development: an evolutionary basis for the aetiology of autism. J Evol Biol 2006; 19(4):1007-1032.
57. Cook Jr EH, Lindgren V, Leventhal BL et al. Autism or atypical autism in maternally but not paternally derived proximal 15q duplication. Am J Hum Genet 1997; 60(4):928-934.
58. Schroer RJ, Phelan MC, Michaelis RC et al. Autism and maternally derived aberrations of chromosome 15q. Am J Med Genet 1998; 76(4):327-336.
59. Repetto GM, White LM, Bader PJ et al. Interstitial duplications of chromosome region 15q11q13: clinical and molecular characterization. Am J Med Genet 1998; 79(2):82-89.

60. Nurmi EL, Dowd M, Tadevosyan-Leyfer O et al. Exploratory subsetting of autism families based on savant skills improves evidence of genetic linkage to 15q11-q13. J Am Acad Child Adolesc Psychiatry 2003; 42(7):856-863.
61. Bittel DC, Kibiryeva N, Talebizadeh Z et al. Microarray analysis of gene/transcript expression in Angelman syndrome: deletion versus UPD. Genomics 2005; 85(1):85-91.
62. Ashley-Koch A, Wolpert CM, Menold MM et al. Genetic studies of autistic disorder and chromosome 7. Genomics 1999; 61(3):227-236.
63. Donnelly SL, Wolpert CM, Menold MM et al. Female with autistic disorder and monosomy X (Turner syndrome): parent-of-origin effect of the X chromosome. Am J Med Genet 2000; 96(3):312-316.
64. Ronald A, Happe F, Plomin R. The genetic relationship between individual differences in social and nonsocial behaviors characteristic of autism. Dev Sci 2005; 8(5):444-458.
65. Ronald A, Happe F, Bolton P et al. Genetic heterogeneity between the three components of the autism spectrum: a twin study. J Am Acad Child Adolesc Psychiatry 2006; 45(6):691-699.
66. Creswell CS, Skuse DH. Autism in association with Turner syndrome: genetic implications for male vulnerability to passive developmental disorders. Neurocase 1999; 5:101-108.
67. Good CD, Lawrence K, Thomas NS et al. Dosage-sensitive X-linked locus influences the development of amygdala and orbitofrontal cortex and fear recognition in humans. Brain 126(Pt 11):2431-2446.
68. Elgar K, Campbell R, Skuse D. Are you looking at me? Accuracy in processing line-of-sight in Turner syndrome. Proc R Soc Lond B Biol Sci 2002; 269(1508):2415-2422.
69. Wicker B, Perrett DI, Baron-Cohen S et al. Being the target of another's emotion: a PET study. Neuropsychologia 2003; 41(2):139-146.
70. Kesler SR, Blasey CM, Brown WE et al. Effects of X-monosomy and X-linked imprinting on superior temporal gyrus morphology in Turner syndrome. Biol Psychiatry 2003; 54(6):636-646.
71. Baron-Cohen S. Mindblindness: An essay on autism and theory of mind. Cambridge, Mass: MIT Press/Bradford Books, 1995.
72. Skuse DH, James RS, Bishop DV et al. Evidence from Turner's syndrome of an imprinted X-linked locus affecting cognitive function. Nature 1997; 387(6634):705-708.
73. Skuse DH. Imprinting, the X-chromosome and the male brain: explaining sex differences in the liability to autism. Pediatr Res 2000; 47(1):9-16.
74. Zechner U, Wilda M, Kehrer-Sawatzki H et al. A high density of X-linked genes for general cognitive ability: a run-away process shaping human evolution? Trends Genet 2001; 17(12):697-701.
75. Raefski AS, O'Neill MJ. Identification of a cluster of X-linked imprinted genes in mice. Nat Genet 2005; 37(6):620-624.
76. Davies W, Isles A, Smith R et al. Xlr3b is a new imprinted candidate for X-linked parent-of-origin effects on cognitive function in mice. Nat Genet 2005; 37(6):625-629.
77. Haig D, Westoby M. Parent specific gene expression and the triploid endosperm. Am Nat 1989; 134:147-155.
78. Haig D. Genomic Imprinting and Kinship. New Brunswick, NJ: Rutgers University Press, 2002.
79. Canteras NS, Chiavegatto S, Valle LE et al. Severe reduction of rat defensive behavior to a predator by discrete hypothalamic chemical lesions. Brain Res Bull 1997; 44(3):297-305.
80. Risold PY, Thompson RH, Swanson LW. The structural organization of connections between hypothalamus and cerebral cortex. Brain Res Rev 1997; 24:197-254.
81. Smuts B. Social relationships and life history of primates. In: Galloway A, Zihlman AL eds. The Evolving Female: A Life-History Perspective. Princeton, NJ: Princeton University Press, 1997:60-68.
82. Cheney D, Seyfarth R, Smuts B. Social relationships and social cognition in nonhuman primates. Science 1986; 234(4782):1361-1366.
83. Shallice T. Specific impairments of planning. Philos Trans R Soc Lond B Biol Sci 1982; 298(1089): 199-209.
84. Dunbar RIM. Neocortex size as a constraint on group size in primates. J Hum Evol 1992; 20:469-493.
85. Joffe TH, Dunbar RI. Visual and socio-cognitive information processing in primate brain evolution. Proc Biol Sci 1997; 264(1386):1303-1307.
86. Seyfarth RM, Cheney DL. What are big brains for? Proc Natl Acad Sci 2002; 99(7):4141-4142.
87. Lefebvre L, Viville S, Barton SC et al. Abnormal maternal behavior and growth retardation associated with loss of the imprinted gene Mest. Nat Genet 1998; 20:163-169.
88. Welch MG, Ruggiero DA. Predicted role of secretin and oxytocin in the treatment of behavioral and developmental disorders: implications for autism. Int Rev Neurobiol 2005; 71:273-315.
89. Wu S, Jia M, Ruan Y et al. Positive association of the oxytocin receptor gene (OXTR) with autism in the Chinese Han population. Biol Psychiatry 2005; 58(1):74-77.
90. Green L, Fein D, Modahl C et al. Oxytocin and autistic disorder: alterations in peptide forms. Biol Psychiatry 2001; 50(8):609-613.

91. Insel TR, O'Brien DJ, Leckman JF. Oxytocin, vasopressin and autism: is there a connection? Biol Psychiatry 1999; 45(2):145-157.
92. Kennedy DP, Redcay E, Courchesne E. Failing to deactivate: resting functional abnormalities in autism. Proc Natl Acad Sci 2006; 103(21):8275-8280.
93. Belmonte MK, Carper RA. Monozygotic twins with Asperger syndrome: differences in behavior reflect variations in brain structure and function. Brain Cogn 2006; 61(1):110-121.
94. Chandana SR, Behen ME, Juhasz C et al. Significance of abnormalities in developmental trajectory and asymmetry of cortical serotonin synthesis in autism. Int J Dev Neurosci 2005; 23(2-3):171-182.
95. Baron-Cohen S. The essential difference: men, women and the extreme male brain. London: Allen Lane Penguin Books, 2003.
96. Mental Illness. What does it mean? In: Great Britain Department of Health Great Britain Department of Health Social Services Inspectorate, ed. Health of the Nation. London, UK: HMSO, 1991.
97. Baron-Cohen S. Two new theories of autism: hyper-systemising and assortative mating. Arch Dis Child 2006; 91(1):2-5.
98. Finkel D, McGue M. Sex differences and non-additivity in heritability of the Multidimensional Personality Questionnaire Scales. J Pers Soc Psychol 1997; 72(4):929-938.
99. Cassidy SB, Morris CA. Behavioral phenotypes in genetic syndromes: genetic clues to human behavior. Adv Pediatr 2002; 49:59-86.
100. Skuse DH. Behavioral phenotypes: what do they teach us? Arch Dis Child 2000; 82(3):222-225.
101. Butler M. Prader-Willi Syndrome: Current understanding of cause and diagnosis. Am J Med Genet 1990; 35:319-332.
102. Moore T, Haig D. Genomic imprinting in mammalian development: a parental tug-of-war. Trends Genet 1991; 7(2):45-49.
103. Friend WC. Psychopathology and nonmendelian inheritance. In: M.V. Seeman, ed(s). Gender and Psychopathology. Washington, DC: American Psychiatric Press, 1995:41-61.
104. Rankinen T, Zuberi A, Chagnon YC et al. The human obesity gene map: the 2005 update. Obesity (Silver Springs) 2006; 14(4):529-644.
105. Dong C, Li WD, Geller F et al. Possible genomic imprinting of three human obesity-related genetic loci. Am J Hum Genet 2005; 76(3):427-437.
106. Lindsay RS, Kobes S, Knowler WC et al. Genome-wide linkage analysis assessing parent-of-origin effects in the inheritance of type 2 diabetes and BMI in Pima Indians. Diabetes 2001; 50(12):2850-2857.
107. Parkinson WL, Weingarten HP. Dissociative analysis of ventromedial hypothalamic obesity syndrome. Am J Physiol 1990; 259(4 Pt 2):R829-835.
108. Tecott LH, Sun LM, Akana SF et al. Eating disorder and epilepsy in mice lacking 5-HT2c serotonin receptors. Nature 1995; 374(6522):542-546.
109. Halford JC, Blundell JE. Separate systems for serotonin and leptin in appetite control. Ann Med 2000; 32(3):222-232.
110. Haig D, Wharton R. Prader-Willi syndrome and the evolution of human childhood. Am J Hum Biol 2003; 15(3):320-329.
111. Bruning JC, Gautam D, Burks DJ et al. Role of brain insulin receptor in control of body weight and reproduction. Science 2000; 289(5487):2122-2125.
112. Bennett ST, Wilson AJ, Esposito L et al. Insulin VNTR allele-specific effect in type 1 diabetes depends on identity of untransmitted paternal allele. The IMDIAB Group. Nat Genet 1997; 17(3):350-352.
113. Le Stunff C, Fallin D, Bougneres P. Paternal transmission of the very common class I INS VNTR alleles predisposes to childhood obesity. Nat Genet 2001; 29(1):96-99.
114. Ong KK, Petry CJ, Barratt BJ et al. Maternal-fetal interactions and birth order influence insulin variable number of tandem repeats allele class associations with head size at birth and childhood weight gain. Diabetes 2004; 53(4):1128-1133.
115. Morison IM, Reeve AE. A catalogue of imprinted genes and parent-of-origin effects in humans and animals. Hum Mol Genet 1998; 7(10):1599-1609.
116. Moore GE, Abu-Amero SN, Bell G et al. Evidence that insulin is imprinted in the human yolk sac. Diabetes 2001; 50(1):199-203.
117. Giddings SJ, King CD, Harman KW et al. Allele specific inactivation of insulin 1 and 2, in the mouse yolk sac, indicates imprinting. Nat Genet 1994; 6(3):310-313.
118. DeChiara TM, Robertson EJ, Efstratiadis A. Parental imprinting of the mouse insulin-like growth factor II gene. Cell 1991; 64(4):849-859.
119. Barlow DP, Stoger R, Herrmann BG et al. The mouse insulin-like growth factor type-2 receptor is imprinted and closely linked to the Tme locus. Nature 1991; 349(6304):84-87.
120. Bartolomei MS, Zemel S, Tilghman SM. Parental imprinting of the mouse H19 gene. Nature 1991; 351(6322):153-155.

121. Bachner-Melman R, Zohar AH, Nemanov L et al. Association between the insulin-like growth factor 2 gene (IGF2) and scores on the Eating Attitudes Test in nonclinical subjects: a family-based study. Am J Psychiatry 2005; 162(12):2256-2262.
122. Cripps RL, Martin-Gronert MS, Ozanne SE. Fetal and perinatal programming of appetite. Clin Sci (Lond) 2005; 109(1):1-11.
123. Isse N, Ogawa Y, Tamura N et al. Structural organization and chromosomal assignment of the human obese gene. J Biol Chem 1995; 270(46):27728-27733.
124. Waterland RA. Do maternal methyl supplements in mice affect DNA methylation of offspring? J Nutr 2003; 133(1):238; author reply 239.
125. Waterland RA, Lin JR, Smith CA et al. Post-weaning diet affects genomic imprinting at the insulin-like growth factor 2 (Igf2) locus. Hum Mol Genet 2006; 15(5):705-716.
126. Gardner DK, Lane M. Ex vivo early embryo development and effects on gene expression and imprinting. Reprod Fertil Dev 2005; 17(3):361-370.
127. Ravelli AC, van Der Meulen JH, Osmond C et al. Obesity at the age of 50 y in men and women exposed to famine prenatally. Am J Clin Nutr 1999; 70(5):811-816.
128. Hales CN, Barker DJ. Type 2 (non-insulin-dependent) diabetes mellitus: the thrifty phenotype hypothesis. Diabetologia 1992; 35(7):595-601.
129. Junien C, Gallou-Kabani C, Vige A et al. Nutritionnal epigenomics: consequences of unbalanced diets on epigenetics processes of programming during lifespan and between generations. Ann Endocrinol (Paris) 2005; 66(2 Pt 3):2S19-2S28.
130. Karno M, Golding JM, Sorenson SB et al. The epidemiology of obsessive-compulsive disorder in five US communities. Arch Gen Psychiatry 1988; 45(12):1094-1099.
131. Alsobrook 2nd JP, Zohar AH, Leboyer M et al. Association between the COMT locus and obsessive-compulsive disorder in females but not males. Am J Med Genet 2002; 114(1):116-120.
132. State MW, Dykens EM, Rosner B et al. Obsessive-compulsive symptoms in Prader-Willi and "Prader-Willi-Like" patients. J Am Acad Child Adolesc Psychiatry 1999; 38(3):329-334.
133. Martin A, State M Anderson GM et al. Cerebrospinal fluid levels of oxytocin in Prader-Willi syndrome: a preliminary report. Biol Psychiatry 1998; 44(12):1349-1352.
134. Leckman JF, Goodman WK, North WG et al. Elevated cerebrospinal fluid levels of oxytocin in obsessive-compulsive disorder. Comparison with Tourette's syndrome and healthy controls. Arch Gen Psychiatry 1994; 51(10):782-792.
135. Walitza S, Wewetzer C, Warnke A et al. 5-HT2A promoter polymorphism -1438G/A in children and adolescents with obsessive-compulsive disorders. Mol Psychiatry 2002; 7(10):1054-1057.
136. Ozaki N, Goldman D, Kaye WH et al. Serotonin transporter missense mutation associated with a complex neuropsychiatric phenotype. Mol Psychiatry 2003; 8(11):895, 933-6.
137. Joel D. Current animal models of obsessive compulsive disorder: A critical review. Prog Neuropsychopharmacol Biol Psychiatry, 2006.
138. Cavaillé J, Buiting K, Kiefmann M et al. Identification of brain-specific and imprinted small nucleolar RNA genes exhibiting an unusual genomic organization. Proc Natl Acad Sci 2000; 97(26):14311-14316.
139. Filipowicz W. Imprinted expression of small nucleolar RNAs in brain: Time for RNomics. Proc Natl Acad Sci 2000; 97(26):14035-14037.
140. de los Santos T, Schweizer J, Rees CA et al. Small evolutionarily conserved RNA, resembling C/D box small nucleolar RNA, is transcribed from pwcr1, a novel imprinted gene in the Prader-Willi deletion region, which Is highly expressed in brain. Am J Hum Genet 2000; 67(5):1067-1082.
141. Swanson JM, Sergeant JA, Taylor E et al. Attention-deficit hyperactivity disorder and hyperkinetic disorder. Lancet 1998; 351(9100):429-433.
142. APA. Diagnostic and Statistical Manual of Mental Disorders. Washington, DC: American Psychiatric Association, 1994.
143. Leo D, Sorrentino E, Volpicelli F et al. Altered midbrain dopaminergic neurotransmission during development in an animal model of ADHD. Neurosci Biobehav Rev 2003; 27(7):661-669.
144. Burke JD, Loeber R, Lahey BB et al. Developmental transitions among affective and behavioral disorders in adolescent boys. J Child Psychol Psychiatry 2005; 46(11):1200-1210.
145. Volk HE, Neuman RJ, Todd RD. A systematic evaluation of ADHD and comorbid psychopathology in a population-based twin sample. J Am Acad Child Adolesc Psychiatry 2005; 44(8):768-775.
146. Dick DM, Viken RJ, Kaprio J et al. Understanding the covariation among childhood externalizing symptoms: genetic and environmental influences on conduct disorder, attention deficit hyperactivity disorder and oppositional defiant disorder symptoms. J Abnorm Child Psychol 2005; 33(2):219-229.
147. Willcutt EG, Pennington BF, Chhabildas NA et al. Psychiatric comorbidity associated with DSM-IV ADHD in a nonreferred sample of twins. J Am Acad Child Adolesc Psychiatry 1999; 38(11):1355-1362.

148. Thapar A, Harrington R, McGuffin P. Examining the comorbidity of ADHD-related behaviors and conduct problems using a twin study design. Br J Psychiatry 2001; 179:224-229.
149. Faraone SV, Biederman J. Do attention deficit hyperactivity disorder and major depression share familial risk factors? J Nerv Ment Dis 1997; 185(9):533-541.
150. Tsai SJ, Cheng CY, Yu YW et al. Association study of a brain-derived neurotrophic-factor genetic polymorphism and major depressive disorders, symptomatology and antidepressant response. Am J Med Genet B Neuropsychiatr Genet 2003; 123(1):19-22.
151. Blackman GL, Ostrander R, Herman KC. Children with ADHD and depression: a multisource, multimethod assessment of clinical, social and academic functioning. J Atten Disord 2005; 8(4):195-207.
152. LeBlanc N, Morin D. Depressive symptoms and associated factors in children with attention deficit hyperactivity disorder. J Child Adol Psychiatric Nurs 2004; 17(2):49-55.
153. Mick E, Biederman J, Santangelo S et al. The influence of gender in the familial association between ADHD and major depression. J Nerv Ment Dis 2003; 191(11):699-705.
154. Faraone SV, Biederman J, Mennin D et al. Attention-deficit hyperactivity disorder with bipolar disorder: a familial subtype? J Am Acad Child Adolesc Psychiatry 1997; 36(10):1378-1387.
155. Biederman J, Faraone SV, Keenan K et al. Evidence of familial association between attention deficit disorder and major affective disorders. Arch Gen Psychiatry 1991; 48(7):633-642.
156. Faraone SV, Biederman J, Mennin D et al. Bipolar and antisocial disorders among relatives of ADHD children: parsing familial subtypes of illness. Am J Med Genet 1998; 81(1):108-116.
157. Banaschewski T, Brandeis D, Heinrich H et al. Association of ADHD and conduct disorder—brain electrical evidence for the existence of a distinct subtype. J Child Psychol Psychiatry 2003; 44(3): 356-376.
158. Doyle AE, Faraone SV. Familial links between attention deficit hyperactivity disorder, conduct disorder and bipolar disorder. Curr Psychiatry Rep 2002; 4(2):146-152.
159. Smalley SL, McGough JJ, Del'Homme M et al. Familial clustering of symptoms and disruptive behaviors in multiplex families with attention-deficit/hyperactivity disorder. J Am Acad Child Adolesc Psychiatry 2000; 39(9):1135-1143.
160. Faraone SV, Biederman J, Monuteaux MC. Attention-deficit disorder and conduct disorder in girls: evidence for a familial subtype. Biol Psychiatry 2000; 48(1):21-29.
161. Wigren M, Hansen S. ADHD symptoms and insistence on sameness in Prader-Willi syndrome. J Intellect Disabil Res 2005; 49(Pt 6):449-456.
162. Couper R. Prader-Willi syndrome. J Paediatr Child Health 1999; 35(4):331-334.
163. Watanabe H, Ohmori O, Abe K. Recurrent brief depression in Prader-Willi syndrome: a case report. Psychiatr Genet 1997; 7(1):41-44.
164. Dykens EM, Cassidy SB. Correlates of maladaptive behavior in children and adults with Prader-Willi syndrome. Am J Med Genet 1995; 60(6):546-549.
165. Swaab DF. Neuropeptides in hypothalamic neuronal disorders. Int Rev Cytol 2004; 240:305-375.
166. Mao R, Jalal SM, Snow K et al. Characteristics of two cases with dup(15)(q11.2-q12): one of maternal and one of paternal origin. Genet Med 2000; 2(2):131-135.
167. Cattanach BM, Kirk M. Differential activity of maternally and paternally derived chromosome regions in mice. Nature 1985; 315(6019):496-498.
168. Lichter DG, Jackson LA, Schachter M. Clinical evidence of genomic imprinting in Tourette's Syndrome. Neurology 1995; 45:924-928.
169. McMahon FJ, Stine OC, Meyers DA et al. Patterns of maternal transmission in bipolar affective disorder. Am J Hum Genet 1995; 56:1277-1286.
170. Borglum AD, Kirov G, Craddock N et al. Possible parent-of-origin effect of Dopa decarboxylase in susceptibility to bipolar affective disorder. Am J Med Genet, B: Neuropsychiatr Genet 2003; 117(1):18-22.
171. Hawi Z, Dring M, Kirley A et al. Serotonergic system and attention deficit hyperactivity disorder (ADHD): a potential susceptibility locus at the 5-HT(1B) receptor gene in 273 nuclear families from a multi-centre sample. Mol Psychiatry 2002; 7(7):718-725.
172. Sheehan K, Lowe N, Kirley A et al. Tryptophan hydroxylase 2 (TPH2) gene variants associated with ADHD. Mol Psychiatry 2005; 10(10):944-949.
173. Curran S, Purcell S, Craig I et al. The serotonin transporter gene as a QTL for ADHD. Am J Med Genet B Neuropsychiatr Genet 2005; 134(1):42-47.
174. Stoltenberg SF, Glass JM, Chermack ST et al. Possible association between response inhibition and a variant in the brain-expressed tryptophan hydroxylase-2 gene. Psychiatr Genet 2006; 16(1):35-38.
175. Zill P, Baghai TC, Zwanzger P et al. SNP and haplotype analysis of a novel tryptophan hydroxylase isoform (TPH2) gene provide evidence for association with major depression. Mol Psychiatry 2004; 9(11):1030-1036.
176. Zill P, Buttner A, Eisenmenger W et al. Single nucleotide polymorphism and haplotype analysis of a novel tryptophan hydroxylase isoform (TPH2) gene in suicide victims. Biol Psychiatry 2004; 56(8):581-586.

177. Koponen E, Rantamaki T, Voikar V et al. Enhanced BDNF signaling is associated with an antidepressant-like behavioral response and changes in brain monoamines. Cell Mol Neurobiol 2005; 25(6):973-980.
178. Mattson MP, Maudsley S, Martin B. BDNF and 5-HT: a dynamic duo in age-related neuronal plasticity and neurodegenerative disorders. Trends Neurosci 2004; 27(10):589-594.
179. Kent L, Green E, Hawi Z et al. Association of the paternally transmitted copy of common Valine allele of the Val66Met polymorphism of the brain-derived neurotrophic factor (BDNF) gene with susceptibility to ADHD. Mol Psychiatry 2005; 10(10):939-943.
180. Goos LM, Ezzatian P, Schachar R. Parent-of-origin effects in Attention Deficit Hyperactivity Disorder (ADHD). Psychiatry Res in press.
181. Marmorstein NR, Iacono WG. Major depression and conduct disorder in youth: associations with parental psychopathology and parent-child conflict. J Child Psychol Psychiatry 2004; 45(2):377-386.
182. Marmorstein NR, Malone SM, Iacono WG. Psychiatric disorders among offspring of depressed mothers: associations with paternal psychopathology. Am J Psychiatry 2004; 161(9):1588-1594.
183. Hicks BM, Krueger RF, Iacono WG et al. Family transmission and heritability of externalizing disorders: a twin-family study. Arch Gen Psychiatry 2004; 61(9):922-928.
184. Pfiffner LJ, McBurnett K, Rathouz PJ et al. Family correlates of oppositional and conduct disorders in children with attention deficit/hyperactivity disorder. J Abnorm Child Psychol 2005; 33(5):551-563.
185. Haber JR, Jacob T, Heath AC. Paternal alcoholism and offspring conduct disorder: evidence for the 'common genes' hypothesis. Twin Res Hum Genet 2005; 8(2):120-131.
186. Comings DE, Gade-Andavolu R, Gonzalez N et al. Comparison of the role of dopamine, serotonin and noradrenaline genes in ADHD, ODD and conduct disorder: multivariate regression analysis of 20 genes. Clin Genet 2000; 57(3):178-196.
187. Castellanos FX, Tannock R. Neuroscience of attention-deficit/hyperactivity disorder: the search for endophenotypes. Nat Rev Neurosci 2002; 3(8):617-628.
188. Crosbie J, Schachar R. Deficient inhibition as a marker for familial ADHD. Am J Psychiatry 2001; 158:1884-1890.
189. Schachar R, Mota VL, Logan GD et al. Confirmation of an inhibitory control deficit in attention-deficit/hyperactivity disorder. J Abnorm Child Psychol 2000; 28(3):227-235.
190. Kruglyak L, Lander ES. High-resolution genetic mapping of complex traits. Am J Hum Genet 1995; 56(5):1212-1223.
191. Gershon ES, Badner JA, Detera-Wadleigh SD et al. Maternal inheritance and chromosome 18 allele sharing in unilineal bipolar illness pedigrees. Am J Med Genet 1996; 67(2):202-207.
192. Biederman J, Newcorn J, Sprich S. Comorbidity of attention deficit hyperactivity disorder with conduct, depressive, anxiety and other disorders. Am J Psychiatry 1991; 148(5):564-577.
193. Faraone SV, Biederman J, Keenan K et al. A family-genetic study of girls with DSM-III attention deficit disorder. Am J Psychiatr 1991; 148(1):112-117.
194. Schmitz M, Cadore L, Paczko M et al. Neuropsychological performance in DSM-IV ADHD subtypes: an exploratory study with untreated adolescents. Can J Psychiatr 2002; 47(9):863-869.
195. Murphy KR, Barkley RA, Bush T. Young adults with attention deficit hyperactivity disorder: subtype differences in comorbidity, educational and clinical history. J Nerv Ment Dis 2002; 190(3):147-157.
196. Chhabildas N, Pennington BF, Willcutt EG. A comparison of the neuropsychological profiles of the DSM-IV subtypes of ADHD. J Abn Child Psychol 2001; 29(6):529-540.
197. Faraone SV, Biederman J, Mick E et al. A family study of psychiatric comorbidity in girls and boys with attention-deficit/hyperactivity disorder. Biol Psychiatry 2001; 50(8):586-592.
198. Biederman J, Mick E, Faraone SV et al. Influence of gender on attention deficit hyperactivity disorder in children referred to a psychiatric clinic. Am J Psychiatr 2002; 159(1):36-42.
199. Graetz BW, Sawyer MG, Baghurst P. Gender differences among children with DSM-IV ADHD in Australia. J Am Acad Child Adolesc Psychiatry 2005; 44(2):159-168.
200. Dykens EM, Hodapp RM. Research in mental retardation: toward an etiologic approach. J Child Psychol Psychiatry 2001; 42(1):49-71.
201. Abdolmaleky HM, Thiagalingam S, Wilcox M. Genetics and epigenetics in major psychiatric disorders: dilemmas, achievements, applications and future scope. Am J Pharmacogenomics 2005; 5(3):149-160.
202. Koenig K, Klin A, Schultz R. Deficits in social attribution ability in Prader–Willi Syndrome. J Autism Dev Disord 2004; 34(5).
203. Dick DM, Edenberg HJ, Xuei X et al. Association of GABRG3 with alcohol dependence. Alcohol Clin Exp Res 2004; 28(1):4-9.
204. Cookson WO, Moffatt MF. Genetics of asthma and allergic disease. Hum Mol Genet 2000; 9(16):2359-2364.
205. Cookson WO, Young RP, Sandford AJ et al. Maternal inheritance of atopic IgE responsiveness on chromosome 11q. Lancet 1992; 340(8816):381-384.

flowering plants (angiosperms) which arose about 140 million years ago, reproduction occurs by double fertilization whereby one of the two sperm cells in the pollen grain fertilizes the egg cell to produce the embryo, while the other sperm cell fertilizes the binucleate central cell to generate the triploid endosperm which nourishes the plant embryo (Fig. 1). In flowering plants, imprinting has so far only been demonstrated for a monocot (maize) and a dicot (Arabidopsis) and because it is expected that imprinting is limited to the triploid endosperm, it is expected that imprinting will not be found in gymnosperms which lack double fertilisation and hence a triploid endosperm.[28]

The conflict hypothesis proposes that selection will drive monoallelic expression of paternally derived alleles that increase maternal resource allocation to the offspring (growth enhancers), while growth inhibitors are predicted to be expressed from the maternally derived allele only.[29] The first two imprinted genes discovered were a pair of genes (*Igf2, Igf2R*) which interact at the protein level, yet had opposite imprinting expression and mutant phenotypes on embryo growth. Experimental support for the parental conflict theory for imprinting evolution is largely derived from observations of expected overgrowth or undergrowth phenotypes when imprinted loci are disrupted in mammals (e.g., *GNASxl, Grb10, IGF2, Igf2R*) or plants (e.g., *MEDEA*).[30-35]

While the parental conflict theory for the evolution of imprinting is the most widely discussed, a wide range of other theories have also been posited regarding why imprinting has arisen.[36,37] These include theories based on prevention of parthenogenesis,[38-40] dosage compensation,[41] meiotic recombination,[42] expression variance minimization,[43] intralocus sexual conflict,[44] maternal-offspring coadaptation[45] and selection for parental similarity.[46]

Amongst these theories, both the meiotic recombination-based and the ovarian timebomb theories have some evidence of experimental support. It has been proposed that there is a mechanistic link between the processes of genomic imprinting and meiotic recombination.[42] In the mid 1990s, it was discovered that imprinted regions of the human genome displayed sex-specific meiotic recombination.[47,48] It has since been demonstrated that imprinted regions of the human genome display high rates of meiotic recombination.[49,50]

The ovarian timebomb theory for the evolution of imprinting proposes that imprinting has evolved to prevent parthenogenesis.[38-40] Support for this theory is derived from the failure of diploid gynogenetic or androgenetic embryos to develop to term.[1,2,7] More recently, the deletion of the H19 imprinted region is considered to have facilitated the generation of the first parthenogenetic mouse, Kaguya.[51,52] In the plant kingdom, where apomictic plants, which reproduce asexually via their seeds, arise naturally in over 400 species, the relative parental contributions in the triploid endosperm of pseudogamous apomicts are often found to mimic the [2 maternal:1 paternal] genome ratio observed in sexually reproducing diploid angiosperms.[53] In addition, screens for autonomous apomixis in the model plants have identified the Fertilisation Independent Seed (FIS) genes (*MEDEA/FIS1, FIS2* and *FIE/FIS3*), two of which (*MEDEA* and *FIS2*) are amongst the four known imprinted genes in Arabidopsis.[54] Nonetheless, the main support for the ovarian timebomb theory is currently derived from studies of imprinted genes and parthenogenesis in mammals.

Genomic Imprinting in Plants

Parent-of-origin effects on angiosperm seed development have been widely described in plants.[56-58] However, since genomic imprinting was first demonstrated in maize,[59] the study of gene-specific imprinting during seed development in Arabidopsis, the plant genetic model, has grown in importance over the past decade.[16,60,61]

Because of the nature of the sexual cycle in angiosperms (Fig. 1), parent-of-origin effects observed on seed development resulting from inter-ploidy crosses[62,63] can be the consequence of a range of mechanisms, of which imprinting is one possibility.[53,55] Because both the ploidy and the relative parental genome contributions differ in triploid seeds generated from reciprocal crosses between diploid and tetraploids, such effects could also result from dosage effects in the endosperm.[64] Also the relative extent of control of the embryo, endosperm and maternal tissues over seed characteristics (e.g., mass) remains not fully elucidated. Hence, it is currently not possible to

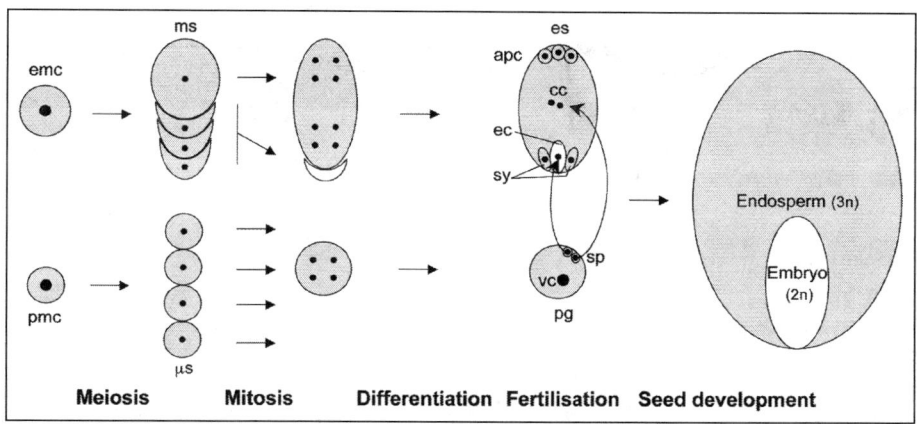

Figure 1. Schematic representation of gametogenesis, fertilisation and seed development in Arabidopsis thaliana. In plants, two identical male gametes (sp) are delivered by the pollen tube to two different female gametes, the egg cell (eg) and the central cell (cc). The process of double fertilisation leads to the development of the diploid embryo and a terminally differentiated triploid tissue, the endosperm. The endosperm is a nourishing tissue connecting the embryo to the sporophytic maternal tissue surrounding the seed. The figure highlights the differences between male and female gametogenesis and the differential contribution of each parent into the progeny. Various genetic mechanisms have been highlighting in causing parent-of-origin effects during seed development[55]:- disproportionate contribution to the endosperm; - plastidic and cytoplasmic inheritance; - gene expression during gametogenesis and; - differential parental allelic expression (i.e., imprinting) in the developing seed. apc: antipodal cells; cc: central cell; ec: egg cell; emc: embryo sac mother cell; es: embryo sac; ms: megaspore; μs: microspore; pg: pollen grain; sp: sperm cells; sy: synergid cells; vc: vegetative cell.

interpret inter-ploidy parent-of-origin phenotypic effects on seed development as solely due to genomic imprinting.[53,55,65]

Early evidence of imprinting in plants at a gene locus was observed in the endosperm of maize where only some alleles (e.g., *R, dzr1 + MO17*) of a gene in specific maize inbred lines behave as imprinted genes.[57,59,66] While this type of allele-specific imprinting is considered to be different to gene-specific imprinting where the majority of alleles at a locus are imprinted,[16,67] it could also be interpreted as a form of imprinting polymorphism similar to that observed for the *Igf2R* and *WT1* genes which are only imprinted in some humans.[16,67]

In contrast to the over 80 known imprinted mammalian genes, gene specific imprinting has so far been demonstrated for ten genes in plant seeds, with six imprinted genes identified in maize, and four identified in *Arabidopsis thaliana* (Table 1). While mammalian imprinted genes are found to be clustered in the fully sequenced human and mouse genomes,[68] it is not possible at present to determine whether imprinted genes in plant genomes exhibit any significant clustering because of the small number of known imprinted genes in plants.

Imprinting Regulation at the Maternally Expressed *MEDEA* Locus in *Arabidopsis thaliana*

The regulation of *MEDEA* (*MEA*) expression during gametophyte and seed development in *Arabidopsis thaliana* is the subject of intensive studies and is used as one of the main models to determine the molecular mechanisms regulating imprinting in plants.

Loss of function mutations of *MEA* (*mea*) show a parent-of-origin maternal effect. When inherited maternally (but not paternally), the mutant *mea* allele induces endosperm and embryo

Table 1. Overview of known gene-specific imprinting genes in plants

Imprinted Gene	Expressed Allele	Species	References
MEDEA	Maternal	Arabidopsis thaliana	68, 73
FWA	Maternal	Arabidopsis thaliana	82
PHE1	Paternal	Arabidopsis thaliana	87
FIS2	Maternal	Arabidopsis thaliana	78
fie1 and fie2	Maternal	Zea mays	89
meg1	Maternal	Zea mays	90
nrp	Maternal	Zea mays	91
peg1	Paternal	Zea mays	92
Mez1	Maternal	Zea mays	93

over proliferation and eventually leads to seed abortion.[32,69] *MEA* encodes a SET-domain protein homologous to Enhancer of Zeste[32,69] and controls seed development as a component of a Polycomb Repressive Complex 2 (PRC2),[70,71] for review, see. refs. 72 and 73.

Evidence that *MEA* is a maternally expressed imprinted gene in Arabidopsis seeds has been firmly established using in situ RNA hybridisation[69,74] and reporter gene fusions.[75]

Because many flowering plants are hermaphrodites, an imprint distinguishing male and female genomes can be deposited during the gametophytic phase of the sexual cycle (Fig. 1). Using in situ hybridisation and reporter gene fusions, *MEA* expression has been identified in the female gametophyte and not in the male gametophyte.[74,75]

Using the power of the genetic model, Arabidopsis, candidate suppressors or enhancers of imprinting at the *MEA* locus have been identified. Such analysis has revealed that *DEMETER* (*DME*), a functional DNA glycosylase which can excise 5-methylcytosines,[76] plays a role in imprinting regulation at the *MEA* locus. *DEMETER* mutants (*dme*) induce a similar maternal-effect seed abortion phenotype as mea, as well as pleiotropic developmental defects.[76]

DME is expressed only before fertilisation in the central cell of the female gametophyte, a precursor of the endosperm (Fig. 1), and regulates *MEA* expression by activating the maternal *MEA* allele. *DME* is not expressed in the male gametophyte. *DME* expression in the female gametophyte is necessary for *MEA* RNA accumulation, imprinting of a MEA::GFP fusion and ectopic expression of *DME* induce ectopic expression of a MEA::GFP fusion.[76] Although *DME* expression decreases rapidly after fertilisation, its effect on *MEA* expression is observed later during the development of the endosperm leading to the proposal that DME modifies a heritable epigenetic mark at the *MEA* locus.[76]

Interestingly, *MET1*, encoding the CpG methyltransferase of maintenance homologous to the mammalian Dmnt1 maintenance methyltransferase, counteracts DME control of *MEA*. Loss of function alleles of *MET1* interact genetically with dme to suppress the seed abortion phenotype[77] and both the *MEA* wild type RNA level and the wild type spatial and temporal expression of the *MEA*::GFP transgene is restored in the double *dme met1* mutant.[77]

The maintenance of paternal MEA silencing does not depend on DNA methylation.[78,79] Indeed, in *met1* and other mutant backgrounds affecting DNA methylation, paternal *MEA* silencing is not released indicating that lack of DNA methylation does not directly cause loss of paternal *MEA* silencing.[78] Moreover, biallelic expression of *MEA* observed in the embryo is triggered from highly methylated alleles[78] showing that DNA methylation is also not sufficient to suppress *MEA* expression.

Instead, maintenance of MEA paternal silencing is dependent on the expression of the *MEA*-containing PRC2 complex.[78,80] The normally silent paternal *MEA* allele is reactivated in homozygous *mea* mutants, indicating that the *MEA*-containing PRC2 autoregulates silencing of the paternal *MEA* allele. *Histone H3 methylation on Lysine 27* (*H3K27*), a target of PcG SET domain and a hallmark of PcG-dependent transcriptional silencing, is reported upstream and downstream of *MEA*.[78,80] Loss of H3K27 methylation in these regions is dependent on MEA PRC2 PcG complex and correlates with loss of paternal *MEA* silencing.[78,80] These regions appear to colocalise with the MET1-dependent DNA methylation regions. While autoregulatory activity of *MEA* maintains the paternal allele silent later during post-fertilisation seed development, the auto-regulation exhibited by *MEA* also extends to the down-regulation of the maternal allele around the time of fertilization.[81] The autorepression of the maternal *MEA* allele has been shown to be direct and independent of the MEA-FIE PcG complex.[81]

Imprinting Regulation at the Maternally Expressed *FWA* Locus in *Arabidopsis thaliana*

The *FWA* gene was the second imprinted gene to be discovered in *Arabidopsis thaliana*,[82,117] where it behaves as a maternally expressed imprinted gene in the endosperm.[83] Using RT-PCR and in situ localisation of a FWA-GFP protein fusion, Kinoshita et al (2004) have shown that *FWA* is expressed in the female gametophyte before fertilisation and in the endosperm only from the maternal genome following fertilisation. *FWA* is silenced in all other plant tissues.

Interestingly, *DME* is also necessary for *FWA* RNA and FWA::GFP protein accumulation in the ovule suggesting that establishment of *FWA* imprinting may be regulated via the same DME/MET1 antagonistic pathway as MEA.[83] But in contrast to *MEA*, maintenance of paternal silencing of *FWA* is dependent on *MET1*. Indeed, when a wt maternal plant is crossed with a met1 paternal plant, *FWA* imprinting in the endosperm is lost suggesting a direct involvement of CpG methylation in the maintenance of *FWA* paternal silencing.[75,83] A MET1-dependent DNA methylation region, composed of two sets of tandem repeats encompassed within a SINE retrotransposon related element, has been detected in the promoter and 5' UTR of *FWA*.[83,84] The maternal copy of this region is hypomethylated in the endosperm[83] and maintenance of silencing of *FWA* in the body of the plant is also associated with DNA methylation and Histone H3 Lysine 9 methylation, a hallmark of silent heterochromatin.[85,86,116]

Imprinting Regulation at the Paternally Expressed *PHE1* Locus in *Arabidopsis thaliana*

In *Arabidopsis thaliana*, the *PHERES1* (*PHE1*) locus encodes a type-I MADs box gene which is a downstream target of the maternally expressed imprinted gene *MEDEA*.[87] The *PHE1* locus was subsequently shown to also be subject to genomic imprinting, albeit as the first paternally expressed imprinted gene to be identified in Arabidopsis.[88] The *PHE1* locus is regulated by histone trimethylation on H3K27 residues mediated by at least two different PcG complexes in plants, containing the SET domain proteins MEDEA or CURLY LEAF/SWINGER.[89]

Imprinting Regulation at the Maternally Expressed *FIS2* Locus in *Arabidopsis thaliana*

The fourth imprinted gene discovered in the Arabidopsis genome is the Fertilisation Independent Seed 2 (*FIS2*) gene which is maternally expressed in the endosperm.[79] At the imprinted *FIS2* locus, the maternal allele of *FIS2* is also activated by DME in the central cell. Paternal silencing at the imprinted *FIS2* locus is dependent on the MET1 maintenance methyltransferase being present in the pollen parent, and could be localised to a small CpG-rich region upstream of *FIS2*.

Differentially Methylated Domains (DMDs) and Imprinting Regulation in Plants

In mammals, there is evidence that some domains (differentially methylated domains, DMDs) proximal to imprinted genes can act as imprinting control regions (ICRs).[3] DMDs may constitute one type of cis-acting element which if present in certain genomic contexts facilitate the imprinting of proximal genes.[95] DMDs associated with some imprinted mammalian genes (e.g., *H19* and *Igf2*) likely act as methylation-sensitive chromatin insulators.[96-98] It is important to stress that not all DMDs are likely to function as ICRs, and all ICRs may not necessarily be DMDs.

In Arabidopsis endosperm, MET1-dependent differentially methylated domains (DMDs) have been detected upstream (-500 bp) and downstream (MEA-ISR region) of *MEA*[77,78,99] and lack of cytosine methylation in these regions, and most particularly in a MEA-ISR region composed of tandem repeats downstream of the gene, correlate with the active state of the maternal *MEA* allele in the endosperm.[77,78] However, both reporter gene studies with *MEA* promoter:GUS fusion constructs that recapitulate imprinting, and the functional analysis of *MEA* imprinting in Arabidopsis accessions that do not contain the tandem downstream repeats, have revealed that the presence of the downstream MEA-ISR repeats is not required for *MEA* imprinting.[100]

In maize, the *FERTILISATION INDEPENDENT ENDOSPERM 1* (*FIE1*) and *FIE2* genes have been identified as maternally expressed imprinted genes in the endosperm.[90] In the embryo, *FIE1* is not expressed, while *FIE2* expression is biallelic. In maize, it was possible to isolate sufficient quantities of male and female gametes to conduct a detailed analysis of methylation patterns at the *FIE1* and *FIE2* loci.[101] Initial expression analysis of *FIE* promoter:reporter gene constructs which recapitulated imprinting for both *FIE1* and *FIE2* indicated the regions of the *FIE1* (−1489 to + 1354) and *FIE2* (−1537 to + 1541) promoters that were necessary for imprinting regulation. Bisulphite sequencing analysis was performed on these regions for endosperm, embryos, sperm cells, egg cell and central cell tissue. For *FIE1*, two DMDs were found where methylation was present in the paternal alleles in endosperm and in the sperm cell, but not present in the maternal alleles in the endosperm and central cell. Methylation at these DMDs was found to be present in the egg cell and the maternal allele in the embryo. The strong correlation between these DMDs and the maternal-specific expression of the *FIE1* locus in the endosperm is suggestive that these DMDs may act as ICRs.

Intriguingly, analysis of the *FIE2* locus indicated that de novo methylation of the *FIE2* DMD may be occurring for the non-expressed paternal *FIE2* in the endosperm, as no methylation was detected in the sperm cell at this region. In mammalian imprinting, de novo methyltransferases (Dnmt3a and Dnmt3b) are essential for imprint resetting during the imprint erasure-resetting-maintenance cycle of imprinting.[102] The plant orthologs of Dnmt3 are proteins of the DOMAINS REARRANGED METHYLTRANSFERASE (DRM) family.[103] Analysis of the imprinted *FWA* locus in a *drm1 drm2* mutant background has indicated that activation of the normally silenced paternal *FWA* allele in endosperm is not dependent on DRM1 and DRM2.[83]

The analysis of methylation at imprinted genes in maize has indicated that the expressed maternal alleles of the *FIE* genes are associated with hypomethylated DMDs while the non-expressed paternal alleles are associated with hypermethylated DMDs. While this study is consistent with the proposal that methylation is the default state for the silent allele at some imprinted loci, it also raises the possibility for de novo establishment of methylated DMDs via an as yet elusive imprint-setting mechanism in plant endosperm.

Emerging Models for Imprinting Regulation in Plants

The results presented above show that in plants, like in mammals,[104] DNA methylation and chromatin structure are key regulatory components of genomic imprinting.[105] To date, genomic imprinting in plants appears to be restricted to the endosperm, a terminally differentiated tissue analogous to the placenta. In contrast to mammals, if imprinting in plants is wholly limited to the endosperm, imprinting does not need to be reset at each sexual generation.[58]

The apparent absence of an active de novo methylation process in plants is further indicated following the re-introduction of wild type methylase activity in a *MET1* mutant. Indeed, DNA remethylation is extremely slow and, even for some loci it can be negligible.[105] Then, it is not surprising that, in contrast to mammals, epigenetic mutations (i.e., epimutations i.e., transgenerational inheritance of epigenetic state) have been frequently identified in plants.[106,107]

Genomic Imprinting in Plants

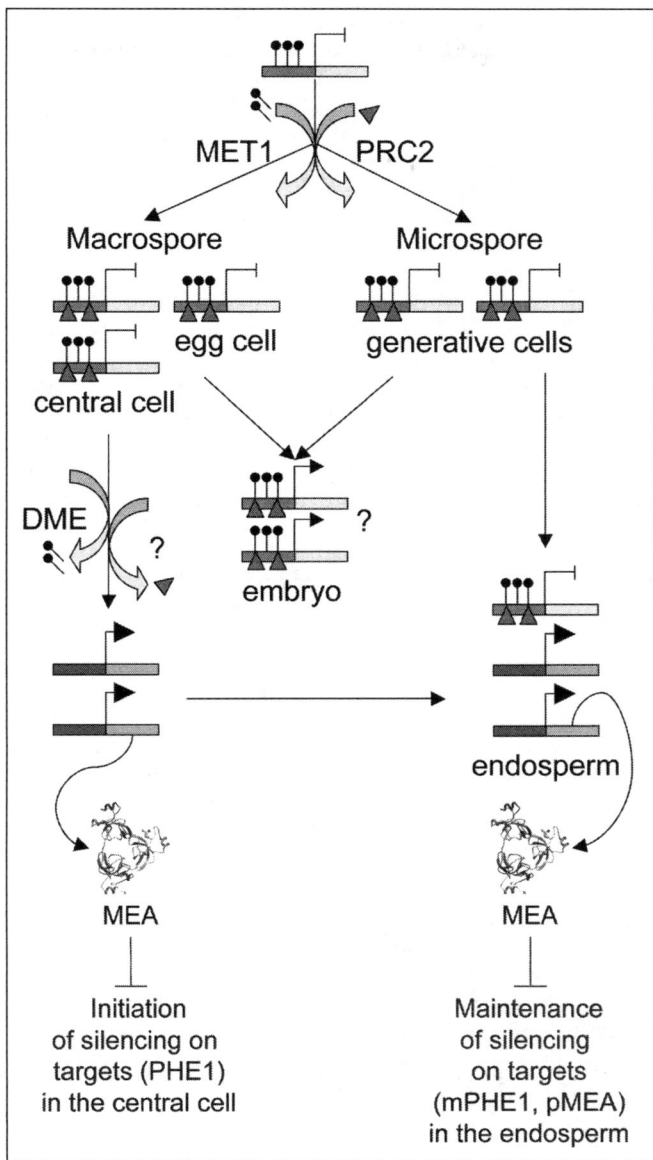

Figure 2. A model of regulation of genomic imprinting in the endosperm. *FIS2, FWA* and *MEA* are imprinted in the endosperm and are subjected to silencing by DNA maintenance methylation (*FWA, FIS2*) by *MET1* or H3K27 methylation by PRC2 (*MEA*). During female gametophytic development, the activity of DME in the central cell removes methylated cytosines and by an unknown mechanism induces other putative epigenetic modifications (such as H3K27 on *MEA*) which in turn lead to the expression of the maternal allele. *MEA* expression in the central cell initiates the repression of additional loci such as the paternally expressed imprinted PHE1 locus. Then, maternal *PHE1* repression as well as paternal *MEA* repression is maintained by a negative feedback loop involving a PRC2 complex. MET1: DNA methyltransferase 1; DME: *DEMETER*; MEA: *MEDEA*; pMEA: *MEA* paternal silencing; PHE1: *PHERES 1*; mPHE1: *PHE1* maternal silencing; PRC2: Polycomb Repressive Complex 2.

Nonetheless, while similar key regulatory components seem to control genomic imprinting in plants and in mammals, fundamentally different models establishing imprinting have been proposed.[77] As the presence of MET1 appears to prevent the activation of *MEA* in a *dme* mutant background, it has been suggested that the essential function of DME in the female gametophyte is to counteract the methyltransferase activity of MET1.[77,83]

Two models have been proposed to explain how DME overcomes MET1 methyltransferase activity and could activate *MEA* or *FWA* in the central cell.[77,83] In one model, the nicking activity of DME could induce local chromatin remodeling influencing DNA methylation and reactivation of target genes.[76,77] However, data showing that Transcriptional Gene Silencing (TGS) is impaired in the DNA repair mutant *rpa2*[108] suggest to the contrary that DNA nicks would instead induce silencing. This would be in agreement with the suspected evolutionary role of gene silencing in plants.[109] In an alternative model, the DNA glycosylase activity of DME could directly removes methylcytosine[77,82] as this has been shown in related mammalian[110] or Arabidopsis DNA glycosylases (Fig. 2).[111] The observation that mutations in the *Arabidopsis DNA glycosylase ROS1* suppress TGS supports this model.[111] In mammals gene silencing through methylation appears to be a major regulatory mechanism for many imprinted loci. However, in plants, activation through de-methylation of key regulatory genes in the female gametophyte appears to initiate imprinting in the endosperm. Thus, methylation and silencing of imprinted alleles in plants is probably a default state.[105]

It is still unclear how silencing has been once initiated during the evolution of the *MEA* and *FWA* loci. The presence of tandem repeats or transposable elements (TE), features common to the peri- and centromeric heterochromatic regions, has been annotated for both loci.[84,100] Tandem repeats and TEs have been implicated in an RNAi pathway loop establishing and maintaining silencing through heterochromatinisation.[86,112] Interestingly, small RNA of ~24 nt corresponding to tandem repeats in both *MEA* and *FWA* loci have been observed in the wild type.[84] However the exact role of the tandem repeats and TE in imprinting is still discussed,[100,113] their existence suggest that RNAi-like pathways could have been essential to establishing DNA methylation and could be essential to maintain DNA methylation and thus direct *DME* activity to specific loci in the central cell. Using RNA-directed-DNA-Methylation (RdDM), the importance and localisation of DNA methylation relative to a gene of interest in the initiation and maintenance of silencing and the initiation of DME-dependent activation could be tested.

While maintenance of the silent FWA paternal allele relies on MET1-dependent DNA methylation,[79,83] maintenance of *MEA* paternal silencing is not MET1 dependent.[78,79] Maintenance of *MEA* paternal silencing necessitates the action of SET domain proteins and is maintained through a feedback loop.[60,80] The MEA PRC2 exerts its direct control on *MEA* and its other targets like *PHERES1* (*PHE1*), a type I MADS-box gene,[87] by parental imprinting, paternal silencing in the case of MEA[78,80] and maternal silencing in the case of *PHE1*.[88] Interestingly, PcG proteins and the antagonist Trithorax Group (TrG) proteins regulate genomic imprinting also in metazoans.[21,114,115] However, *MEA* regulation by a feedback loop is a unique example of a PcG gene regulating its own imprinting.

The study of imprinting in plants has significantly grown over the past decade. The wide availability of mutants affecting epigenetic processes in plants has allowed the plant imprinting field to apply such mutants to genetically dissect the mechanisms controlling imprinting in plants. While the number of imprinted plant genes remains small by comparison to the mammalian imprinting field, the discovery of new imprinted genes and mechanisms (e.g., DNA glycosylation) will allow a more comparative approach to understanding the independent evolution of imprinting in the mammalian and plant kingdoms.

Acknowledgements

OG, CS and SLD acknowledge the support of Science Foundation Ireland.

References

1. Barton SC, Surani MA, Norris ML. Role of paternal and maternal genomes in mouse development. Nature 1984; 311:374-376.
2. McGrath J, Solter D. Completion of mouse embryogenesis requires both the maternal and paternal genomes. Cell 1984; 37:179-183.
3. Reik W, Walter J. Genomic imprinting: parental influence on the genome. Nat Rev Genet 2001; 2:21-32.
4. Stern C. The nucleus and somatic cell variation. J Cell Physiol 1958; 52:1-27; discussion 27-34.
5. Crouse HV. The nature of the influence of x-translocations on sex of progeny in Sciara coprophila. Chromosoma 1960; 11:146-166.
6. Goday C, Esteban MR. Chromosome elimination in sciarid flies. Bioessays 2001; 23:242-250.
7. Surani MA, Barton SC, Norris ML. Development of reconstituted mouse eggs suggests imprinting of the genome during gametogenesis. Nature 1984; 308:548-550.
8. de la Casa-Esperon E, Sapienza C. Natural selection and the evolution of genome imprinting. Annu Rev Genet 2003; 37:349-370.
9. Wolffe AP, Matzke MA. Epigenetics: regulation through repression. Science 1999; 286:481-486.
10. Wu C, Morris JR. Genes, genetics, and epigenetics: a correspondence. Science 2001; 293:1103-1105.
11. Henikoff S, McKittrick E, Ahmad K. Epigenetics, histone H3 variants, the inheritance of chromatin states. Cold Spring Harb Symp Quant Biol 2004; 69:235-243.
12. Tilghman SM. The sins of the fathers and mothers: genomic imprinting in mammalian development. Cell 1999; 96:185-193.
13. Walter J, Paulsen M. Imprinting and disease. Semin Cell Dev Biol 2003; 14:101-110.
14. Morison IM, Ramsay JP, Spencer HG. A census of mammalian imprinting. Trends Genet 2005; 21:457-465.
15. Moore T. Genetic conflict, genomic imprinting and establishment of the epigenotype in relation to growth. Reproduction 2001; 122:185-193.
16. Scott RJ, Spielman M. Genomic imprinting in plants and mammals: how life history constrains convergence. Cytogenet Genome Res 2006b; 113:53-67.
17. Haig D, Westoby M. Parent specific gene expression and the triploid endosperm. American Naturalist 1989; 134:147-155.
18. Moore T, Haig D. Genomic imprinting in mammalian development: a parental tug-of-war. Trends Genet 1991; 7:45-49.
19. Lloyd VK, Sinclair DA, Grigliatti TA. Genomic imprinting and position-effect variegation in Drosophila melanogaster. Genetics 1999; 151:1503-1516.
20. Lloyd V. Parental imprinting in Drosophila. Genetica 2000; 109:35-44.
21. Joanis V, Lloyd VK. Genomic imprinting in Drosophila is maintained by the products of Suppressor of variegation and trithorax group, but not Polycomb group, genes. Mol Genet Genomics 2002; 268:103-112.
22. Killian JK, Byrd JC, Jirtle JV et al. M6P/IGF2R imprinting evolution in mammals. Mol Cell 2000; 5:707-716.
23. O'Neill MJ, Ingram RS, Vrana PB et al. Allelic expression of IGF2 in marsupials and birds. Dev Genes Evol 2000; 210:18-20.
24. Killian JK, Nolan CM, Wylie AA et al. Divergent evolution in M6P/IGF2R imprinting from the Jurassic to the Quaternary. Hum Mol Genet 2001; 10:1721-1728.
25. Hahn Y, Yang SK, Chung JH. Structure and expression of the zebrafish mest gene, an ortholog of mammalian imprinted gene PEG1/MEST. Biochim Biophys Acta 2005; 1731:125-132.
26. Colosi DC, Martin D, More K et al. Genomic organization and allelic expression of UBE3A in chicken. Gene 2006; 383:93-98.
27. Lawton BR, Sevigny L, Obergfell C et al. Allelic expression of IGF2 in live-bearing, matrotrophic fishes. Dev Genes Evol 2005; 215:207-212.
28. Baroux C, Spillane C, Grossniklaus U. Evolutionary origins of the endosperm in flowering plants. Genome Biol 2002a; 3, reviews 1026.
29. Haig D. Genomic imprinting and kinship: how good is the evidence? Annu Rev Genet 2004; 38:553-585.
30. Barlow DP, Stoger R, Herrmann BG et al. The mouse insulin-like growth factor type-2 receptor is imprinted and closely linked to the Tme locus. Nature 1991; 349:84-87.
31. DeChiara TM, Robertson EJ, Efstratiadis A. Parental imprinting of the mouse insulin-like growth factor II gene. Cell 1991; 64:849-859.
32. Grossniklaus U, Vielle-Calzada JP, Hoeppner MA et al. Maternal control of embryogenesis by MEDEA, a polycomb group gene in Arabidopsis. Science 1998; 280:446-450.
33. Constancia M, Hemberger M, Hughes J et al. Placental-specific IGF-II is a major modulator of placental and fetal growth. Nature 2002; 417:945-948.

34. Charalambous M, Smith FM, Bennett WR et al. Disruption of the imprinted Grb10 gene leads to disproportionate overgrowth by an Igf2-independent mechanism. Proc Natl Acad Sci USA 2003; 100:8292-8297.
35. Plagge A, Gordon E, Dean W et al. The imprinted signaling protein XL alpha s is required for postnatal adaptation to feeding. Nat Genet 2004; 36:818-826.
36. Hurst LD, McVean GT. Growth effects of uniparental disomies and the conflict theory of genomic imprinting. Trends Genet 1997; 13:436-443.
37. Hurst LD, McVean GT. Do we understand the evolution of genomic imprinting? Curr Opin Genet Dev 1998; 8:701-708.
38. Solter D. Differential imprinting and expression of maternal and paternal genomes. Annu Rev Genet 1988; 22:127-146.
39. Solter D. Refusing the ovarian time bomb. Trends Genet 1994; 10, 346; author reply 348-349.
40. Varmuza S, Mann M. Genomic imprinting—defusing the ovarian time bomb. Trends Genet 1994; 10:118-123.
41. Iwasa Y. The conflict theory of genomic imprinting: how much can be explained? Curr Top Dev Biol 1998; 40:255-293.
42. Pardo-Manuel de Villena F, de la Casa-Esperon E, Sapienza C. Natural selection and the function of genome imprinting: beyond the silenced minority. Trends Genet 2000; 16:573-579.
43. Weisstein AE, Spencer HG. The evolution of genomic imprinting via variance minimization: an evolutionary genetic model. Genetics 2003; 165:205-222.
44. Day T, Bonduriansky R. Intralocus sexual conflict can drive the evolution of genomic imprinting. Genetics 2004; 167:1537-1546.
45. Wolf JB, Hager R. A maternal-offspring coadaptation theory for the evolution of genomic imprinting. PLoS Biol 2006; 4, e380.
46. Spencer HG, Clark AG. A chip off the old block: a model for the evolution of genomic imprinting via selection for parental similarity. Genetics 2006; 174:931-935.
47. Paldi A, Gyapay G, Jami J. Imprinted chromosomal regions of the human genome display sex-specific meiotic recombination frequencies. Curr Biol 1995; 5:1030-1035.
48. Robinson WP, Lalande M. Sex-specific meiotic recombination in the Prader--Willi/Angelman syndrome imprinted region. Hum Mol Genet 1995; 4:801-806.
49. Lercher MJ, Hurst LD. Imprinted chromosomal regions of the human genome have unusually high recombination rates. Genetics 2003; 165:1629-1632.
50. Sandovici I, Kassovska-Bratinova S, Vaughan JE et al. Human imprinted chromosomal regions are historical hot-spots of recombination. PLoS Genet 2006; 2, e101.
51. Kono T, Obata Y, Wu Q et al. Birth of parthenogenetic mice that can develop to adulthood. Nature 2004; 428:860-864.
52. Kono T. Genomic imprinting is a barrier to parthenogenesis in mammals. Cytogenet Genome Res 2006; 113:31-35.
53. Grossniklaus U, Spillane C, Page DR et al. Genomic imprinting and seed development: endosperm formation with and without sex. Curr Opin Plant Biol 2001; 4:21-27.
54. Scott RJ, Spielman M. Deeper into the maize: new insights into genomic imprinting in plants. Bioessays 2006a; 28:1167-1171.
55. Dilkes BP, Comai L. A differential dosage hypothesis for parental effects in seed development. Plant Cell 2004; 16:3174-3180.
56. Kermicle JL, Alleman M. Gametic imprinting in maize in relation to the angiosperm life cycle. Dev 1990; Suppl, 9-14.
57. Messing J, Grossniklaus U. Genomic imprinting in plants. Results Probl Cell Differ 1999; 25:23-40.
58. Alleman M, Doctor J. Genomic imprinting in plants: observations and evolutionary implications. Plant Mol Biol 2000; 43:147-161.
59. Kermicle J. Dependence of the R-mottled aleurone phenotype in maize on mode of sexual transmission. Genetics 1970; 66:69-85.
60. Gehring M, Choi Y, Fischer RL. Imprinting and seed development. Plant Cell 2004; 16 Suppl, S203-213.
61. Kohler C, Grossniklaus U. Seed development and genomic imprinting in plants. Prog Mol Subcell Biol 2005; 38:237-262.
62. Scott RJ, Spielman M, Bailey J et al. Parent-of-origin effects on seed development in Arabidopsis thaliana. Development 1998; 125:3329-3341.
63. Adams S, Vinkenoog R, Spielman M et al. Parent-of-origin effects on seed development in Arabidopsis thaliana require DNA methylation. Development 2000; 127:2493-2502.
64. Birchler JA. Dosage analysis of maize endosperm development. Annu Rev Genet 1993; 27:181-204.

65. von Wangenheim KH, Peterson HP. Aberrant endosperm development in interploidy crosses reveals a timer of differentiation. Dev Biol 2004; 270:277-289.
66. Chaudhuri S, Messing J. Allele-specific parental imprinting of dzr1, a posttranscriptional regulator of zein accumulation. Proc Natl Acad Sci USA 1994; 91:4867-4871.
67. Baroux C, Spillane C, Grossniklaus U. Genomic imprinting during seed development. Adv Genet 2002b; 46:165-214.
68. Verona RI, Mann MR, Bartolomei MS. Genomic imprinting: intricacies of epigenetic regulation in clusters. Annu Rev Cell Dev Biol 2003; 19:237-259.
69. Kinoshita T, Yadegari R, Harada JJ et al. Imprinting of the MEDEA polycomb gene in the Arabidopsis endosperm. Plant Cell 1999; 11:1945-1952.
70. Spillane C, MacDougall C, Stock C et al. Interaction of the Arabidopsis polycomb group proteins FIE and MEA mediates their common phenotypes. Curr Biol 2000; 10:1535-1538.
71. Yadegari R, Kinoshita T, Lotan O et al. Mutations in the FIE and MEA genes that encode interacting polycomb proteins cause parent-of-origin effects on seed development by distinct mechanisms. Plant Cell 2000; 12:2367-2382.
72. Sorensen MB, Chaudhury AM, Robert H et al. Polycomb group genes control pattern formation in plant seed. Curr Biol 2001; 11:277-281.
73. Guitton AE, Berger F. Control of reproduction by Polycomb Group complexes in animals and plants. Int J Dev Biol 2005; 49:707-716.
74. Vielle-Calzada JP, Thomas J, Spillane C et al. Maintenance of genomic imprinting at the Arabidopsis medea locus requires zygotic DDM1 activity. Genes Dev 1999; 13:2971-2982.
75. Luo M, Bilodeau P, Dennis ES et al. Expression and parent-of-origin effects for FIS2, MEA, FIE in the endosperm and embryo of developing Arabidopsis seeds. Proc Natl Acad Sci USA 2000; 97:10637-10642.
76. Choi Y, Gehring M, Johnson L et al. DEMETER, a DNA glycosylase domain protein, is required for endosperm gene imprinting and seed viability in arabidopsis. Cell 2002; 110:33-42.
77. Xiao W, Gehring M, Choi Y et al. Imprinting of the MEA Polycomb gene is controlled by antagonism between MET1 methyltransferase and DME glycosylase. Dev Cell 2003; 5:891-901.
78. Gehring M, Huh JH, Hsieh TF et al. DEMETER DNA glycosylase establishes MEDEA polycomb gene self-imprinting by allele-specific demethylation. Cell 2006; 124:495-506.
79. Jullien PE, Kinoshita T, Ohad N et al. Maintenance of DNA methylation during the Arabidopsis life cycle is essential for parental imprinting. Plant Cell 2006a; 18:1360-1372.
80. Jullien PE, Katz A, Oliva M et al. Polycomb group complexes self-regulate imprinting of the Polycomb group gene MEDEA in Arabidopsis. Curr Biol 2006b; 16:486-492.
81. Baroux C, Gagliardini V, Page DR et al. Dynamic regulatory interactions of Polycomb group genes: MEDEA autoregulation is required for imprinted gene expression in Arabidopsis. Genes Dev 2006; 20:1081-1086.
82. Kakutani T, Kato M, Kinoshita T et al. Control of development and transposon movement by DNA methylation in Arabidopsis thaliana. Cold Spring Harb Symp Quant Biol 2004; 69:139-143.
83. Kinoshita T, Miura A, Choi Y et al. One-way control of FWA imprinting in Arabidopsis endosperm by DNA methylation. Science 2004; 303:521-523.
84. Lippman Z, Gendrel AV, Black M et al. Role of transposable elements in heterochromatin and epigenetic control. Nature 2004; 430:471-476.
85. Soppe WJ, Jasencakova Z, Houben A et al. DNA methylation controls histone H3 lysine 9 methylation and heterochromatin assembly in Arabidopsis. EMBO J 2002; 21:6549-6559.
86. Lippman Z, Martienssen R. The role of RNA interference in heterochromatic silencing. Nature 2004; 431:364-370.
87. Kohler C, Hennig L, Spillane C et al. The Polycomb-group protein MEDEA regulates seed development by controlling expression of the MADS-box gene PHERES1. Genes Dev 2003; 17:1540-1553.
88. Kohler C, Page DR, Gagliardini V et al. The Arabidopsis thaliana MEDEA Polycomb group protein controls expression of PHERES1 by parental imprinting. Nat Genet 2005; 37:28-30.
89. Makarevich G, Leroy O, Akinci U et al. Different Polycomb group complexes regulate common target genes in Arabidopsis. EMBO Rep 2006; 7:947-952.
90. Danilevskaya ON, Hermon P, Hantke S et al. Duplicated fie genes in maize: expression pattern and imprinting suggest distinct functions. Plant Cell 2003; 15:425-438.
91. Gutierrez-Marcos JF, Costa LM, Biderre-Petit C et al. Maternally expressed gene1 is a novel maize endosperm transfer cell-specific gene with a maternal parent-of-origin pattern of expression. Plant Cell 2004; 16:1288-1301.
92. Guo M, Rupe MA, Danilevskaya ON et al. Genome-wide mRNA profiling reveals heterochronic allelic variation and a new imprinted gene in hybrid maize endosperm. Plant J 2003; 36:30-44.

93. Gutierrez-Marcos JF, Pennington PD, Costa LM et al. Imprinting in the endosperm: a possible role in preventing wide hybridization. Philos Trans R Soc Lond B Biol Sci 2003; 358:1105-1111.
94. Haun W, Laouielle-Duprat S, O'Connell M et al. Genomic imprinting, methylation and molecular evolution of maize Enhancer of zeste (Mez) homologs. Plant Journal in press. 2006.
95. Ginjala V, Holmgren C, Ulleras E et al. Multiple cis elements within the Igf2/H19 insulator domain organize a distance-dependent silencer. A cautionary note. J Biol Chem 2002; 277:5707-5710.
96. Bell AC, West AG, Felsenfeld G. Insulators and boundaries: versatile regulatory elements in the eukaryotic. Science 2001; 291:447-450.
97. Holmgren C, Kanduri C, Dell G et al. CpG methylation regulates the Igf2/H19 insulator. Curr Biol 2001; 11:1128-1130.
98. West AG, Gaszner M, Felsenfeld G. Insulators: many functions, many mechanisms. Genes Dev 2002; 16:271-288.
99. Zilberman D, Cao X, Jacobsen SE. ARGONAUTE4 control of locus-specific siRNA accumulation and DNA and histone methylation. Science 2003; 299:716-719.
100. Spillane C, Baroux C, Escobar-Restrepo JM et al. Transposons and tandem repeats are not involved in the control of genomic imprinting at the MEDEA locus in Arabidopsis. Cold Spring Harb Symp Quant Biol 2004; 69:465-475.
101. Gutierrez-Marcos JF, Costa LM, Dal Pra M et al. Epigenetic asymmetry of imprinted genes in plant gametes. Nat Genet 2006; 38:876-878.
102. Morgan HD, Santos F, Green K et al. Epigenetic reprogramming in mammals. Hum Mol Genet 2005; 14 Spec No 1, R47-58.
103. Chan SW, Henderson IR, Jacobsen SE. Gardening the genome: DNA methylation in Arabidopsis thaliana. Nat Rev Genet 2005; 6:351-360.
104. Reik W, Dean W, Walter J. Epigenetic reprogramming in mammalian development. Science 2001; 293:1089-1093.
105. Takeda S, Paszkowski J. DNA methylation and epigenetic inheritance during plant gametogenesis. Chromosoma, 2006; 115:27-35.
106. Finnegan EJ. Epialleles—a source of random variation in times of stress. Curr Opin Plant Biol 2002; 5:101-106.
107. Kakutani T. Epi-alleles in plants: inheritance of epigenetic information over generations. Plant Cell Physiol 2002; 43:1106-1111.
108. Elmayan T, Proux F, Vaucheret H. Arabidopsis RPA2: A Genetic link among transcriptional gene silencing, DNA repair, DNA replication. Curr Biol 2005; 15:1919-1925.
109. Matzke MA, Aufsatz W, Kanno T et al. Homology-dependent gene silencing and host defense in plants. Adv Genet 2002; 46:235-275.
110. Jost JP, Oakeley EJ, Zhu B et al. 5-Methylcytosine DNA glycosylase participates in the genome-wide loss of DNA methylation occurring during mouse myoblast differentiation. Nucleic Acids Res 2001; 29:4452-4461.
111. Gong Z, Morales-Ruiz T, Ariza RR et al. ROS1, a repressor of transcriptional gene silencing in Arabidopsis, encodes a DNA glycosylase/lyase. Cell 2002; 111:803-814.
112. Gendrel AV, Colot V. Arabidopsis epigenetics: when RNA meets chromatin. Curr Opin Plant Biol 2005; 8:142-147.
113. Martienssen R, Lippman Z, May B et al. Transposons, tandem repeats, the silencing of imprinted genes. Cold Spring Harb Symp Quant Biol 2004; 69:371-379.
114. Wang J, Mager J, Chen Y et al. Imprinted X inactivation maintained by a mouse Polycomb group gene. Nat Genet 2001; 28:371-375.
115. Mager J, Montgomery ND, de Villena FP et al. Genome imprinting regulated by the mouse Polycomb group protein Eed. Nat Genet 2003; 33:502-507.
116. Kakutani T. Genetic characterization of late-flowering traits induced by DNA hypomethylation mutation in Arabidopsis thaliana. Plant J 1997; 12:1447-1451.
117. Soppe WJ, Jacobsen SE, Alonso-Blanco C et al. The late flowering phenotype of fwa mutants is caused by gain-of-function epigenetic alleles of a homeodom ain gene. Mol Cell 2000; 6:791-802.

CHAPTER 8

Imprinted Genes and Human Disease:
An Evolutionary Perspective

Francisco Úbeda* and Jon F. Wilkins

Abstract

Imprinted genes have been associated with a wide range of diseases. Many of these diseases have symptoms that can be understood in the context of the evolutionary forces that favored imprinted expression at these loci. Modulation of perinatal growth and resource acquisition has played a central role in the evolution of imprinting and many of the diseases associated with imprinted genes involve some sort of growth or feeding disorder. In the first part of this chapter, we discuss the relationship between the evolution of imprinting and the clinical manifestations of imprinting-associated diseases. In the second half, we consider the variety of processes that can disrupt imprinted gene expression and function. We ask specifically if there is reason to believe that imprinted genes are particularly susceptible to deregulation—and whether a disruption of an imprinted gene is more likely to have deleterious consequences than a disruption of an unimprinted gene.

There is more to a gene than its DNA sequence. C. H. Waddington used the term "epigenetic" to describe biological differences between tissues that result from the process of development.[1,2] Waddington needed a new term to describe this variation which was neither the result of genotypic differences between the cells nor well described as phenotypic variation. We now understand that heritable modifications of the DNA—such as cytosine methylation—and aspects of chromatin structure—including histone modifications—are the mechanisms underlying what Waddington called the "epigenotype." Epigenetic modifications are established in particular cell lines during development and are responsible for the patterns of gene expression seen in different tissue types.

In contemporary usage, the term epigenetic refers to heritable changes in gene expression that are not coded in the DNA sequence itself.[3] In recent years, much attention has been paid to a particular type of epigenetic variation: genomic imprinting. In the case of imprinting, the maternally and paternally inherited genes within a single cell have epigenetic differences that result in divergent patterns of gene expression.[4] In the simplest scenario, only one of the two alleles at an imprinted locus is expressed. In other cases, an imprinted locus can include a variety of maternally expressed, paternally expressed and biallelically expressed transcripts.[5-10] Some of these transcripts produce different proteins through alternate splicing, while others produce noncoding RNA transcripts.[11-15] Genomic imprinting can also interact with the "epigenotype" in Waddington's sense: many genes are imprinted in a tissue-specific manner, with monoallelic expression in some cell types and biallelic expression in others.[16-20]

Other chapters in this volume cover our current understanding of the mechanisms of imprinting, the phenotypic effects of imprinted genes in mammals and what we know about imprinting in plants. In this chapter we discuss the link between imprinted genes and human disease. First, we consider the phenotypes associated with imprinted genes and ask whether the disorders as-

*Corresponding Author: Francisco Úbeda—St. John's College and Oxford Centre for Gene Function, Oxford University, Oxford OX1 3JP, U.K. Email: francisco.ubeda@st-johns.oxford.ac.uk

Genomic Imprinting, edited by Jon F. Wilkins. ©2008 Landes Bioscience and Springer Science+Business Media.

sociated with these genes share a common motif. Second, we consider the nature and frequency of mutations of imprinted genes. We ask whether we should expect that imprinted genes are particularly fragile. That is, are they more likely to undergo mutation and/or are mutations of imprinted genes particularly likely to result in human disease? In general we consider how the field of evolutionary medicine—the use of evolution to understand why our body's design allows for the existence of disease at all[21]—might contribute to our comprehension of disorders linked to genomic imprinting.

Do Disorders Linked to Imprinted Genes Share a Common Motif?

Many disorders linked to imprinted genes (see Table 1) are related to growth.[22] The Kinship Theory of Imprinting[23,24] explains why genetic loci that influence growth (and particularly the allocation of maternal resources) are prone to evolving imprinted gene expression. However, not all of these diseases are obviously growth related. In some cases, it might be possible to reconcile these disease phenotypes with the more general version of the Kinship Theory. In other cases, these disorders might be related to the mechanism of imprinting, rather than the gene function responsible for the evolution of imprinted expression.

According to the Kinship Theory, the pattern of expression shown by imprinted genes is a consequence of an evolutionary conflict between the maternally inherited (MI) and paternally inherited (PI) alleles at a locus. The theory relies on the notion of the inclusive fitness of an allele,[25] which includes not only the fitness of the individual carrying the allele, but also the fitnesses of other, related individuals who may have inherited an identical copy of that allele. That is, natural selection favors those alleles that maximize the number of copies passed on to future generations, regardless of whether those copies are passed on directly, or though the reproductive success of one's kin. Which other individuals qualify as "relatives" can differ for the MI and PI alleles at a locus. In fact, in an outbred population, the only individuals to whom my MI and PI alleles are equally related are my direct descendants, my full siblings and their direct descendants.

Natural selection favors strategies that increase an allele's inclusive fitness. When the gene affects the fitness of individuals to whom the MI and PI alleles have different degrees of relatedness, an allele's optimal expression strategy will depend on its parental origin.[26] This can lead to silencing of the allele favoring the lower expression level and expression of the other allele at the level that maximizes its inclusive fitness.[27] For example, consider a locus at which an increase in level of expression (which we denote by X) enhances the fitness of the individual carrying the gene, but reduces the fitness of that individual's matrilineal kin (relatives to whom one is related through one's mother), henceforth referred to as resource enhancer. The level of expression that maximizes the inclusive fitness of the PI allele, $\hat{X}_P$, will be higher than that maximizing the inclusive fitness of the MI allele, $\hat{X}_m$. That is, $\hat{X}_P > \hat{X}_m$. Any intermediate level of expression $\hat{X}_P > X > \hat{X}_m$ results in conflict between the MI and PI alleles. If the locus becomes imprinted (acquires the ability to independently regulate the expression level of the MI and PI alleles) this conflict will result in the silencing of the MI allele. Expression of the PI allele will evolve to $\hat{X}_P$, the level that maximizes the patrilineal inclusive fitness. Analogous results apply to a locus where increasing the level of expression, Y, benefits matrilineal kin at the expense of the individual (a resource inhibitor). In this case, however, it is the PI allele that becomes silenced.[28,29]

Most work on imprinted genes has focused on their effects on fetal growth. In this context, the relatedness asymmetries between the maternally and paternally derived alleles are well understood. A gene that enhances fetal growth places a resource demand on the mother, presumably reducing the availability of resources for her other offspring. The magnitude of this fetal demand will be limited by the fact that the MI alleles in the fetus have a fifty per cent chance of being inherited by any one of those other offspring and excessive demand could actually reduce the allele's inclusive fitness (even while increasing the fitness of the individual offspring). Because the mother's other offspring may have a different father, the PI alleles in the fetus care less than the maternally derived alleles about the consequences of increasing resource demand.

The taxonomic and functional distribution of imprinted genes suggest that conflicts over maternal resources have played an especially important role in the evolution of imprinting. Many imprinted genes have been associated with prenatal growth effects.[30] Furthermore, mammalian imprinting appears first to have evolved in the common ancestor of marsupials and eutherian (placental) mammals, coinciding with the origination of viviparity.[31,32] Viviparity and particularly the placental interface, provides an opportunity for the offspring to actively manipulate the availability of maternal resources. In oviparous (egg-laying) species, the mother has unilateral control over the distribution of resources among her offspring. While an intragenomic conflict might, in principle, exist within these offspring, there is no arena in which this conflict can play out.

Genomic imprinting in plants is not yet as well understood, but appears to follow a similar pattern: imprinting has evolved independently in angiosperms (flowering plants) where offspring (seed/fruit) develop in physical contact with the maternal parent. As in the case of mammals, the imprinted genes of angiosperms appear to modulate an offspring's access to maternal resources (see ref. 33 and the chapter by Spillane et al). When the tools of molecular genetics are applied to other plant groups, such a ferns and mosses, we might expect to find a similar set of phenomena.

There is obviously a strong correlation between prenatal growth effects and imprinting. However, similar reasoning applies to any trait where changes in gene expression affect the fitness of matrilineal and patrilineal kin differently.[26] In fact, many imprinted genes have effects on behavior that are difficult to interpret as straightforward extensions of parental conflict. Some of these behavioral effects include maternal care, reactivity to novel environments and social behaviors.[5,34-38] Similarly, viviparity alone is not sufficient to drive the evolution of imprinting. Many viviparous species lack imprinting, including many species of fish.[39]

Three features of mammalian pregnancy are likely responsible for its central role in the evolution of mammalian imprinting. First, there is a large asymmetry of parental resource contribution (maternal, but not paternal, pregnancy). Second, through the placental interface, the offspring plays an active role in soliciting maternal resources. Third, viviparity appears to have been maintained consistently in mammals since its introduction (in contrast to viviparity in other vertebrates, which is more evolutionarily labile).[40-42]

The existence of an inclusive-fitness asymmetry is not unique to mammalian pregnancy. In fact, there may be no single locus in any (biparental) organism for which the optimal expression pattern for the MI and PI alleles are exactly identical. The difference in mammalian pregnancy (and some plant reproductive systems) is a quantitative one. These systems have evolved many imprinted genes because the inclusive-fitness asymmetry is large. Furthermore, the systems are relatively easy to manipulate and have persisted in something like their present form for many millions of years.

Growth and Resource Acquisition

Our discussion of growth-related disorders in pregnancy follows that of Haig.[43] Many of these disorders likely involve the action of imprinted genes (see Table 1), but they should not necessarily be viewed as a consequence of imprinted gene expression. In addition to the conflict between the MI and PI alleles in the offspring, pregnancy is characterized by parent-offspring conflict.[44] The same sorts of inclusive-fitness considerations that underlie the evolutionary explanation for imprinting suggest that the fetus should favor a higher degree of resource demand than the mother. This reasoning applies even in the absence of imprinting. Of course, the set of genes most centrally involved in this conflict should significantly overlap with the set of genes most prone to evolving imprinted gene expression.

While the existence of growth-related disorders does not rely on imprinting, the existence of imprinting might be expected to exacerbate these disorders. In the absence of imprinting, the conflict will be between the maternal interests, on the one hand and the fetal interests (some average of the interests of the MI and PI alleles) on the other. When a growth enhancer becomes imprinted, the MI allele is transcriptionally silenced. At this locus, the parent-offspring conflict then shifts: on one side we still have the maternal interests; on the other, we now have the interests of the PI alleles, which favor a higher level of resource demand than do the fetal genes taken as a

Table 1. Diseases linked to imprinted genes

Disorder	Phenotype	Genes	
Growth and resource acquisition			
Beckwith-Wiedemann Syndrome	Excessively large organs; Fetal and postnatal overgrowth; Low blood sugar in the newborn; Predisposition to tumors	IGF2 (LoF K) LIT1	CDKN1C H19
Growth Related Defects and Metabolic Abnormalities	Placental, pre and post-natal growth excess or deficiency	ARHI IGF2 (LoF D) PEG1 (LoF K) INS (LoF D)	ESX1L (LoF K) GNAS1 H19 (LoI) TSSC3 (LoF K)
Hyperinsulinism	High insulin levels in blood; Resistance to insulin resulting in its overproduction to compensate		
Pre-eclampsia	Elevated blood pressure and sometimes protein in the urine during pregnancy; Swelling of the face and hands		STOX1 (LoF)
Pseudohypoparathyroidism 1A	Lack of response to parathyroid hormone; Low calcium and high phosphate levels in blood; Round face, short stature and hand bones		GNAS1 (LoF K)
Silver-Russell Syndrome	Pre and post natal growth retardation; Predisposition to developmental and motor delays as well as learning disabilities; Body asymmetry	SGCE (LoF)	
Transient Neonatal Diabetes	Inability to use blood glucose for energy resulting in hyperglycemia	PLAGL	LOT1, HYMAI
Post-natal behaviour			
Angelman Syndrome	Feeding problems; Noticeable developmental delays; Pronounced speech impairment; Hyperactivity; Severe movement and balance disorders; Bouts of laughter		UBE3A (LoF K) ATP10C
Reduced Maternal Nursing	Reduced post-natal maternal care. Growth related defects	PEG3 (LoF K)	

continued on next page

Table 1. Continued

Disorder	Phenotype	Genes
Prader-Willi Syndrome	Obesity; short stature; decreased muscle tone; hypogonadism; decrease mental capacity	NDN SNRPN PWCR1 (LoF D) IPW MAGEL2 MKRN3 (LoF K)
Cancer		
Adrenal Cortical Carcinoma		H19 (LoI)
Breast Cancer		PEG1 (LoI) PLAGL1
Hepatoblastoma		H19 (LoI)
Hyatidiform Mole	Uncontrolled growth of the tissue that is supposed to develop into the placenta; Often, there is no fetus at all	
Hyperplasia	Increased cell production in normal tissue or an organ	IGF2 (ΔE)
Nonfunctioning Pituitary Adenomas	Pituitary tumor that does not result in the enhanced production of pituitary hormones	PLAGL1
Retinoblastoma	Retina cancer occurring in children	
Wilms' Tumor	Kidney cancer occurring in children	IGF2 (LoI) NNAT (ΔE) H19 (LoI) CDKN1C

continued on next page

Table 1. Continued

Disorder	Phenotype	Genes
Other		
Autism	Impaired social interactions; Impaired verbal and nonverbal communication; restricted and repetitive patterns of behavior	ATP10C
Bipolar Affective Disorder	Mood swings from mania (exaggerated feeling of well-being, stimulation and grandiosity) to depression (overwhelming feelings of sadness, anxiety and low self-worth)	
McCune-Albright Syndrome	Premature puberty mainly in girls; abnormal fibrous development in the bone; Café-au-lait spots on the skin	GNAS1
Myoclonus Dystonia Syndrome	Obsessive compulsive disorder; panic attacks	SGCE (LoF)
Schizophrenia	Severe problems with thoughts, feelings, behavior, and use of language; delusions often paranoid and persecutory in nature; hallucinations	
Williams-Beuren Syndrome	Mild mental retardation; problems with calcium balance; blood vessels defects	

We indicate the chromosome or chromosomal region involved, whether it is a maternal or paternal UPD and a sketch of the clinical phenotype in each case. (Sources: Imprinted Gene Catalogue; MedlinePlus; Online Mendelian Inheritance in Man).

whole. One of the consequences of imprinting may be an intensification of the pre-existing conflict between mother and fetus.

One of the phenotypes associated with the fetal manipulation of maternal resources is placental invasion. The placenta comprises a fetal portion derived from trophoblasts and a maternal portion derived from the inner layer of the uterine wall. Placental trophoblasts modify maternal arteries to allow greater blood flow through the intervillous space. The greater the penetration of arterial modification into the myometrium, the greater the blood flow and maternal resource transfer to offspring.[43] The higher-than-normal concentration of Insulin-like growth factor type 2 (IGF2)—encoded by the paternally expressed *IGF2* gene—in invasive trophoblasts suggests that IGF2 may influence the extent of placental invasion.[43]

Genetic conflict has also been related to deregulation of maternal blood pressure.[43] The higher the maternal blood pressure the greater the blood flow through the intervillous space and the transfer of resources to the offspring. Paternally inherited genes would favor greater gestational hypertension than their maternally inherited homologs. One of the most common complications of pregnancy (fatal in developing countries) is pregnancy-induced hypertension and its extreme form pre-eclampsia. Pre-eclampsia can be caused by mutations at one of several loci, at least one of which is known to be imprinted—the maternally expressed *STOX1* gene.[45-47] While the exact role of *STOX1* remains unclear, the Kinship Theory would predict that increased expression of *STOX1* would reduce maternal blood pressure.

Increasing the flow of blood to the placenta is one mechanism of fetal resource acquisition. A second is to increase the concentration of nutrients in the maternal circulation. After each meal, maternal insulin prompts the uptake of glucose by maternal cells. During pregnancy, the placenta antagonizes the action of insulin by secreting human placental lactogen (hPL) into the mother's system. This placental hPL generates resistance to insulin in the maternal cells, thereby elevating the level of glucose in the maternal circulation. This manipulation may be the cause of gestational diabetes, which occurs late in pregnancy, but generally resolves quickly following delivery. Imprinting of a locus involved in the placental regulation of hPL could exacerbate this effect and potentially increase the frequency or severity of disorders such as gestational diabetes.

Post-Natal Behavior

After birth, mammals continue to rely heavily on maternal resources (breast milk and supplemental food), although they are no longer transmitted by means of the placenta. The conflict between mother and offspring—and between the PI and MI alleles in the offspring—then shifts primarily into the behavioral arena. Genes expressed in the brain will be under selection to maximize their inclusive fitness, just as they were during pregnancy. However, in this behavioral context, it is often much more difficult to understand the nature of the inclusive-fitness asymmetries that underlie imprinting. Two other chapters in this volume (by Goos and Ragsdale and by Davies et al) focus specifically on behavioral effects associated with imprinted genes and we will discuss the topic only briefly here.

Some of the postnatal behavioral effects of imprinted genes are easily interpreted as the natural extension of the prenatal conflict over maternal resources, such as those affecting suckling and weaning behaviors.[43] In this context, the MI alleles would be expected to favor weaning at an earlier age than the PI alleles. Similarly, PI alleles are expected to more strongly favor behaviors that elicit maternal care. This reasoning has been invoked to explain at least some aspects of the phenotype of two disorders associated with different parental inheritance of deletions or mutations on the long arm of chromosome 15: paternally inherited Prader-Willi syndrome (PWS) and maternally inherited Angelman syndrome (AS).[48]

Each of these disorders exhibits a complex phenotype. AS is associated with enhanced activity, prolonged but poorly coordinated suckling, bouts of laughter, sleeping problems and developmental disorders (speech impairment, movement and balance disorders). Prior to weaning, PWS is associated with reduced activity, poor suckling, weak cry, sleepiness and decreased mental capacity. Following weaning, the child develops an insatiable appetite and becomes obese.[48] Viewed in the light of the

Kinship Theory, the increase in the duration of suckling found in AS may result from the loss of MI alleles that have been selected to reduce the demand for maternal resources. Conversely, PWS is associated with poor food uptake prior to weaning and ravenous food uptake after. The pre-weaning phenotype may result from the loss of PI alleles that been selected to increase the demand of maternal resources. The post-weaning phenotype is more difficult to explain but still is consistent with the Kinship Theory if (a) the offspring's voracious appetite is satisfied primarily by its own foraging efforts and translates in a reduced consumption of breast milk,[48] and (b) the paternal contribution to resource provisioning increases after weaning (F. Úbeda, manuscript in preparation).

Two imprinted genes, *MEST/Peg1* and *Peg3*, show paternal-specific expression in the brains of adult mice. These genes appear to affect the quality of care that mothers provide to their offspring, as knockouts of these genes result in defects in maternal behaviors of nest building, pup retrieval and placentophagy.[37,38] Although the phenotype involves the provisioning of maternal resources to offspring, the conflict in this case is between the mother's two alleles, rather than those of the offspring. The source of this conflict is not obvious, however, since each of the mother's alleles has an equal chance of being passed to each of her offspring. However, if there is some inbreeding in the population (the mother mates with a related male), the offspring could inherit an allele from the father that is identical to one of the mother's alleles.

Under several plausible patterns of inbreeding, the allele inherited from the father is more likely to be identical to the mother's paternally derived allele than to her maternally derived allele. For example, if the mother mates with her own father (i.e., the father and maternal grandfather are the same individual), the mother's paternally derived allele will be more closely related to her own offspring than her maternally derived allele will. If the pattern of inbreeding changes over the course of the female's life, an intra-genomic conflict will arise over the distribution of maternal resources to present and future litters.[49] The conditions under which this selective force might exist are fairly general, but a test of whether this is the factor that is actually responsible for imprinting of these "maternal-care" loci will require close study of multiple species with different patterns of inbreeding.

Cancer

There is mounting evidence that somatic mutations at imprinted loci are associated with a variety of cancers.[50,51] The silencing of one of the alleles turns the imprinted locus functionally haploid. It has been argued that the functional haploidy might increase the risk of cancer by exposing the phenotypic consequences of deleterious recessive mutations. As we discuss in the following section, there are good reasons to believe that deleterious mutations at an imprinted locus are less likely to be recessive than deleterious mutations at other loci (see also the chapter by Moore and Mills).

While we doubt that functional haploidy is the reason for the association of imprinted genes with cancer, there are other features of imprinting that may be relevant. Many genes have evolved imprinting due to a role in modulating fetal growth.[30,52] It is not surprising that many of these same genes influence mitogenic activities in adult somatic tissues. In this context, we suspect that it is not the fact of imprinting that makes these genes associated with tumor growth. Likewise, we doubt that the tumor-suppressing activity of imprinted genes is directly responsible for the evolution of imprinting (but see ref. 53).

Rather, there are a set of genes that affect growth and cell division; this set of genes is more likely to become subject to imprinting and is more likely to be associated with tumorigenic mutations. Imprinting also allows the evolution of antagonistic growth suppressors. Once these exist, they may take on a tumor suppressor role, essentially reducing the selection on other tumor suppressor mechanisms. Here, functional haploidy may be relevant, but not for the obvious reason. In this case, imprinting may replace a biallelically expressed tumor suppressor with a monoallelcally expressed tumor suppressor at a different locus. The result may be a system that is less robust to somatic mutation.

Are Imprinted Genes Particularly Fragile?

There are two reasons why an imprinted gene might be either more likely to express a mutant (sick) phenotype or more susceptible to mutations. First, as metioned above, imprinted genes are functionally haploid. While a recessive mutation has no phenotypic consequences on unimprinted genes, it is exposed on imprinted genes. Second, the expression of imprinted genes, being conditioned by epigenetic factors, is susceptible not only to mutations but also to epimutations. Interestingly, epimutations can be influenced by the environment and do not need to be transient; they might revert after one generation, after a few generations or otherwise become permanent. This opens up a wide range of mutational possibilities.

Mutations

We will distinguish "mutation" (a change in the DNA sequence) from "epimutation" (a heritable change not coded in the DNA sequence). At an unimprinted autosomal locus, loss-of-function mutations are often recessive—a single functional copy of the gene is sufficient to maintain an approximately normal phenotype. At an imprinted locus, one of the two copies is transcriptionally silent and the loss-of-function phenotype depends not on dominance, but on parental origin. A mutation on the silenced allele will have no phenotypic effect. A loss-of-function mutation on the active copy will be equivalent to a homozygous knockout in the absence of imprinting.

Consequently, deleterious recessive mutations, which would have no phenotypic consequences when heterozygous at an unimprinted locus, may have severe phenotypic and fitness effects at an imprinted one. More specifically, the mutant phenotype will be fully revealed half of the time (see Fig. 1). If we assume that most deleterious mutations are recessive, this suggests the functional haploidy associated with genomic imprinting introduces fragility, by increasing the phenotypic and fitness effects of deleterious mutations.

However, there is reason to question the assumption that deleterious mutations of imprinted genes would predominantly be recessive (although this may be true for the genome as a whole). Consider a loss-of-function mutation at an unimprinted locus. This results in a fifty percent reduction in the expression level of the gene. If this deleterious mutation is recessive—the phenotype of a heterozygous carrier of this allele is identical to that of the wild-type—we can infer one of two things. Either there is little or no phenotypic consequence to a fifty percent reduction in gene expression, or there are regulatory feedback mechanisms in place that increase expression from the wild-type allele to compensate for this reduction.

In either case, we should not expect this locus to become imprinted. If a two-fold reduction in expression has no phenotypic consequence, there is little opportunity for an allele to gain an inclusive fitness benefit through changes in the expression level. Similarly, if appropriate feedback mechanisms exist, there will be no selective benefit to silencing the allele favoring lower expression, since the other allele will maintain the overall expression from the locus at a constant level. In fact, the loci where imprinted gene expression will most easily evolve are those loci for which some phenotype is very sensitive to changes in gene expression and where the expression level for each allele is set independently of the total level of gene product. The lack of feedback mechanism might bias imprinted genes towards those whose gene products are exported from the cell (such as growth factors and hormones), making intracellular feedback difficult. The sensitivity to dosage changes implies that deleterious mutations at these loci will not be recessive.

Therefore, even if most deleterious mutations in the genome are recessive, imprinted loci might be more fragile than unimprinted ones not because of their functional haploidy but because of intrinsic properties of the genes that are likely to evolve imprinting, namely, deleterious mutations tend not to be recessive more often than mutations at other loci. Within the group of ancestral genes where imprinting evolved, the functional haploidy of imprinting might actually provide a fitness advantage, by reducing the number of sick phenotypes by one half.

Genomic imprinting is caused by conflict between the two alleles at a single locus, but the outcome of this conflict—silencing of one of the two alleles—creates the potential for conflict among distinct loci.[28,29] Genes with antagonistic effects can become oppositely imprinted and then engage in an arms race characterized by increased expression from both loci. After this escalation occurs, a mutation in either one of the loci can produce a large phenotypic effect. For example, consider an imprinted locus, paternally expressed, with level of expression X and an antagonistic imprinted locus, maternally expressed, with level of expression Y. Assume the difference $X-Y$ determines the

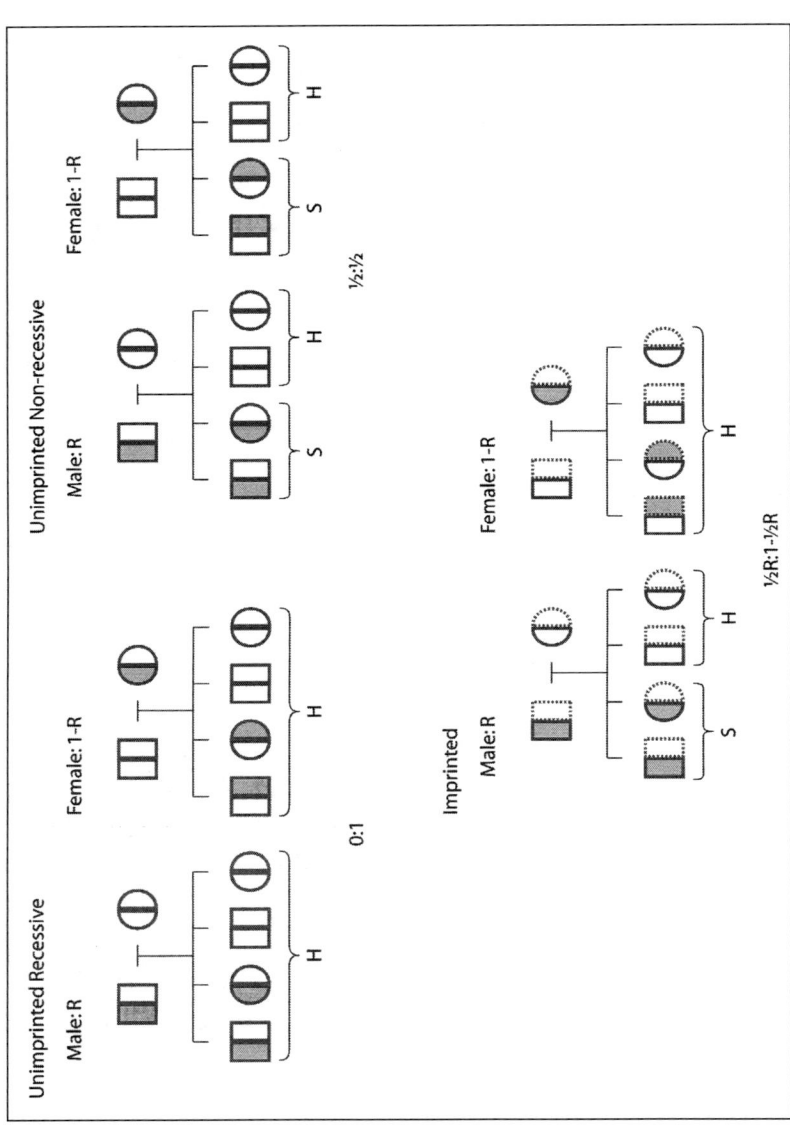

Figure 1. Expected proportion of descendants manifesting a mutant phenotype. We consider a loss-of-function mutation (in grey) in a population with a fraction R of males and $1-R$ of females. In each scenario we consider that the mutation arise in males (squares) and females (circles). The clinical phenotype, healthy (H) or sick (S), is indicated underneath together with its fraction in the population. The scenarios considered are recessive and non-recessive mutations at an unimprinted locus and mutation at an imprinted locus.

blood levels of certain hormone α. Suppose that prior to the acquisition of imprinting at either locus, the expression levels were $X = 2$ and $Y = 1$, resulting in a circulating hormone level of α = 1. Now assume that these two loci evolve antagonistically, resulting in increased expression from each locus, say, $X = 20$ and $Y = 19$. In this example the normal phenotype is unchanged, since we still have α = 1. However, the consequences of a loss-of-function mutation at either locus has been dramatically enhanced. In the original case, deletion of the maternally expressed gene resulted in an increase of α from 1 to 2. After the antagonistic coevolution, deletion of the same gene would increase α from 1 to 20.

Epimutations

Imprinted gene expression is associated with differential epigenetic modifications on each of the chromosomes. These modificaitons include DNA methylation (on cytosines in CpG dinucleotides), as well as histone modifications (including methylation and acetylation). These modifications are established during gametogenesis and are often remodeled following fertilization. Molecular mechanisms exist that reproduce these modifications in the wake of DNA replication.[54] Further epigenetic reprogramming can occur in particular cell lineages during development, resulting in tissue-specific patterns of imprinting. In many cases imprinted gene expression involves the transcription of noncoding RNA transcripts that suppress the production of other transcripts in *cis*.[3,22]

Any of these epigenetic systems is potentially susceptible to failure, resulting in phenotypes that bear a resemblance to genetic disorders, but are not associated with mutations in the DNA sequence. In principle, this sort of failure could occur at any point in the life cycle, including reprogramming errors in gametogenesis or early development, or maintenance errors in particular somatic cell lineages.

DNA methylation patterns are reproduced following DNA replication through the action of a particular DNA methyltransferase (DNMT1). In principle, this provides a passive mechanism by which an epigenetic state, once established, can be maintained across multiple cell divisions. Similarly, it appears that there are passive mechanisms to propagate aspects of chromatin structure, including patterns of histone modification. There is some evidence to suggest that these two systems interact to stabilize epigenetic marks against stochastic loss.[54]

Failure to propagate these marks can result in silencing or reactivation of a particular allele. Of course, susceptibility to somatic epimutations is not limited to imprinted genes. However, as in the case of DNA sequence mutations, the functional haploidy of imprinted loci may make these epimutations particularly detrimental. For instance, inappropriate silencing of the active allele at an imprinted locus will be equivalent to a homozygous loss-of-function mutation in the DNA sequence.

We can distinguish between epimutations that are meiotically heritable and those that are not.[55] Meiotically heritable epimutations occur in the germline and are passed on to the offspring. In some cases (e.g., an error in germline reprogramming at an imprinted locus), we expect the epimutation to persist for a single generation. In other cases, these germline epimutations might be more stable, persisting for several generations, or even being assimilated into the genotype.[56-58]

There is speculation that some of these meiotically heritable mutations may actually respond adaptively to environmental conditions (e.g., nutrients and environmental contaminants), creating a form of trans-generational phenotypic plasticity.[55-58] Dietary conditions at critical ontogenic stages may result in a shortage or an excess of methyl donors.[56,57,59] As a consequence, particular DNA and/or histone methylations might be lost at particular loci, possibly altering their expression pattern. While this mechanism might provide an adaptive response to certain nutrient deficiencies during early development, there may be maladaptive consequences later in life—for instance, as the result of a change in environmental conditions or due to pleiotropic effects of the deregulated imprinted genes.

One candidate example of an environmentally driven epimutation with trans-generational consequences is the so called "metabolic syndrome."[59] Mothers who experience nutritional constraints during pregnancy often have descendants who suffer from glucose and insulin metabolism disorders, weight problems, hypertension, diabetes and cardiovascular diseases.[59-61] Interestingly, these health problems are not limited to the mother's children, but rather can persist in future generations of descendants.[59] The "Fetal Programming Hypothesis of Adult Disease"[22] proposes that dysregulation of imprinted genes may play a role in the clinical phenotype of patients experiencing the metabolic syndorme.[59]

"Metabolic syndrome" is one of many disorders thought to result from epimutations. It is possible that this represents an adaptive response to a nutritional deficiency early in development and that it develops into a disease only when that nutritional constraint is removed. This type of epigenetic response might allow greater adaptability to environmental changes. Epimutations occur frequently (10^{-2} in Tobacco plants[58]) and have a duration of effect that may lend them to certain types of environmental variation. However, adaptive epimutation also has limitations. In particular, the developmental window during which the organism can assimilate environmental cues may be narrow, but the response to those cues may be long lasting. This may lead to maladaptive responses, particularly in the context of contemporary human cultures.

Uniparental Disomies

Genomic imprinting gives rise to the possibility of another type of hereditary defect, uniparental disomies (UPDs). Many animals are sensitive to gene dosage effects and changes in chromosome copy number (e.g., monosomies and trisomies) can often have deleterious effects. For chromosomes containing one or more imprinted loci, parental origin can be as significant as copy number. For instance, an individual who inherits two paternally derived copies of a particular chromosome will have normal gene function at unimprinted loci. However, at a paternally silenced imprinted locus, the individual will functionally be a homozygous knockout. Similarly, each maternally silenced imprinted locus on the chromosome will be expressed at twice its normal level. Most imprinted genes appear to occur in clusters, so that a UPD will typically affect multiple imprinted genes. Not surprisingly, most UPDs are associated with growth abnormalities (see Table 2 and ref. 62). However, given the magnitude of the developmental perturbation typically associated with this type of chromosomal abnormality, it can be difficult to interpret the resulting phenotypes in an evolutionary context.[63-65]

The consequences of imprinting can also be seen in certain trisomies. The deleterious effects of trisomy are not fully attributable to imprinting, but the trisomic phenotype can vary systematically depending on parental origin. For instance, whole-genome triploidy can result in partial hydatidiform moles. While these moles do not go on to form viable offspring, they do undergo partial development and tissue differentiation. Triploid zygotes with an extra paternal genome produce large placentas and small heads. Conversely, zygotes with an extra maternal genome produce small placentas and large heads.[43]

Implications for the Prevention and Treatment of Human Disease

The extension of evolutionary medicine to encompass epigenetic phenomena may prove valuable in the analysis, prevention and treatment of diseases associated with deregulation of imprinted genes. Perhaps the most general insight provided by the evolutionary analysis of imprinted genes is that natural selection does not necessarily act to optimize the fitness (or the health) of an individual organism. Genomic imprinting represents a case where selection to increase inclusive fitness can actually work to the detriment of the individual.[66]

More specifically, the establishment, propagation and interpretation of the epigenetic marks at imprinted loci involve a complex set of mechanisms. Failure of any one of these mechanisms can result in a disease phenotype. In this sense, imprinted genes may represent particularly large mutational targets. Additionally, the escalatory conflicts to which imprinted genes are prone may generate conditions in which mutations (or epimutations) are particularly deleterious.

The reliance of these epigenetic mechanisms on chemical modifications (such as methylation) generates specific nutritional requirements. A deficiency in these or other nutrients can trigger epigenetic reprogramming of particular loci. In some cases, the reprogrammed marks may persist

Table 2. Uniparental disomies

Chromosome	UPD	Phenotype
5 U2AF1RS1	Maternal	Growth retardation
	Paternal	Growth enhancement
6	Maternal	Embryonic lethality; intra-uterine growth retardation
	Paternal	Transient neonatal diabetes mellitus. Growth retardation
7	Maternal	Silver-Russell syndrome (severe intrauterine growth restriction)
7 Grb10	Maternal	Growth retardation
	Paternal	Growth enhancement
11	Paternal	Beckwith-Wiedemann syndrome (fetal and postnatal overgrowth and low blood sugar in the newborn)
14	Maternal	Intra-uterine growth retardation; hypotonia, motor delay and precocious puberty
	Paternal	Growth retardation
15	Maternal	Prader-Willi syndrome (obesity, short stature, decreased muscle tone)
	Paternal	Angelman syndrome (feeding problems, noticeable developmental delays, hyperactivity)
16	Maternal	Intra-uterine growth retardation

We consider four categories: (a) disorders related to growth and resource acquisition (b) disorders related to post-natal behaviour (c) cancers and (d) other disorders. In each case we indicate the disorder, a sketch of the clinical phenotype, the imprinted genes involved (paternally expressed genes in first column and maternally expressed genes in second column). (Sources: Imprinted Gene Catalogue, MedlinePlus; Online Mendelian Inheritance in Man).

across one or more generations. While these changes could simply be a passive byproduct of certain nutrient deficiencies, it is also possible that they represent an adaptive response to environmental cues that are presented in early development. This insight may have implications for the treatment of nutrient deficiencies. In cases like these, simply supplying the missing nutrient at a later developmental stage may create a new set of disease conditions. Nutritional supplementation may have to be coupled with restoration of the original epigenetic state of the modified genes.

Diseases caused by mutations or epimutations at imprinted loci make intriguing candidates for gene therapy. In particular, clinically useable tools for activating or inactivating alleles could provide treatment for many of the disorders mentioned in this chapter. A loss-of-function mutation of the active allele at an imprinted locus might be treated by reactivation of the silent copy. A loss-of-imprinting mutation (inappropriate reactivation of the silenced copy) could be treated through downregulation of the locus as a whole.[3] However, given the complex patterns of regulation and expression at many imprinted loci, this approach may prove technically challenging, is not without potential dangers. Unintended consequences such as a predisposition for tumor formation, will be a danger of any therapy that attempts to quantitatively modify the expression level of imprinted genes.[3] Furthermore, while an intervention of this sort might be beneficial for the patient, the possibility exists that induced epigenetic changes could be passed on to offspring.

Acknowledgements

Wolf Reik and David Haig provided valuable comments on the manuscript. The work of F.U. was supported by a Junior Research Fellowship from St John's College at Oxford University and additional support was provided by Peter Donnelly.

References

1. Waddington CH. An Introduction to Modern Genetics. London: Allen and Unwin; 1939.
2. Waddington CH. The epigenotype. Endeavour 1942; 1:18-20.
3. Egger G, Liang GN, Aparicio A et al. Epigenetics in human disease and prospects for epigenetic therapy. Nature 2004; 429(6990):457-463.
4. Reik W, Walter J. Genomic imprinting: Parental influence on the genome. Nat Rev Genet 2001; 2(1):21-32.
5. Plagge A, Kelsey G. Imprinting the Gnas locus. Cytogenet Genome Res 2006; 113(1-4):178-187.
6. Holmes R, Williamson C, Peters J et al. A comprehensive transcript map of the mouse Gnas imprinted complex. Genome Res 2003; 13(6B):1410-1415.
7. Wroe SF, Kelsey C, Skinner JA et al. An imprinted transcript, antisense to Nesp, adds complexity to the cluster of imprinted genes at the mouse Gnas locus. Proc Natl Acad Sci U S A 2000; 97(7):3342-3346.
8. Weiss U, Ischia R, Eder S et al. Neuroendocrine secretory protein 55 (NESP55): Alternative splicing onto transcripts of the GNAS gene and posttranslational processing of a maternally expressed protein. Neuroendocrinology 2000; 71(3):177-186.
9. Peters J, Wroe SF, Wells CA et al. A cluster of oppositely imprinted transcripts at the Gnas locus in the distal imprinting region of mouse chromosome 2. Proc Natl Acad Sci U S A 1999; 96(7):3830-3835.
10. Hayward BE, Kamiya M, Strain L et al. The human GNAS1 gene is imprinted and encodes distinct paternally and biallelically expressed G proteins. Proc Natl Acad Sci U S A 1998; 95(17):10038-10043.
11. Tufarelli C. The silence RNA keeps: cis mechanisms of RNA mediated epigenetic silencing in mammals. Philos Trans R Soc Lond B Biol Sci 2006; 361(1465):67-79.
12. O'Neill MJ. The influence of noncoding RNAs on allele-specific gene expression in mammals. Hum Mol Genet 2005; 14:R113-R120.
13. Turner M, Williamson C, Nottingham W et al. Nespas: the emerging story of its function in the gnas imprinting cluster. Genet Res 2004; 84(2):118.
14. Hernandez A, Martinez ME, Croteau W et al. Complex organization and structure of sense and antisense transcripts expressed from the DIO3 gene imprinted locus. Genomics 2004; 83(3):413-424.
15. Sleutels F, Zwart R, Barlow DP. The noncoding Air RNA is required for silencing autosomal imprinted genes. Nature 2002; 415(6873):810-813.
16. Liu J, Chen M, Deng CX et al. Identification of the control region for tissue-specific imprinting of the stimulatory G protein alpha-subunit. Proc Natl Acad Sci U S A 2005; 102(15):5513-5518.
17. Williamson CM, Ball ST, Nottingham WT et al. A cis-acting control region is required exclusively for the tissue-specific imprinting of Gnas. Nat Genet 2004; 36(8):894-899.
18. Kashiwagi A, Meguro M, Hoshiya H et al. Predominant maternal expression of the mouse Atp10c in hippocampus and olfactory bulb. J Hum Genet 2003; 48(4):194-198.
19. Yamasaki Y, Kayashima T, Soejima H et al. Neuron-specific relaxation of Igf2r imprinting is associated with neuron-specific histone modifications and lack of its antisense transcript Air. Hum Mol Genet 2005; 14(17):2511-2520.
20. Yamasaki K, Joh K, Ohta T et al. Neurons but not glial cells show reciprocal imprinting of sense and antisense transcripts of Ube3a. Hum Mol Genet 2003; 12(8):837-847.
21. Nesse RM, Williams GC. Why we Get Sick. New York: Times Books; 1994.
22. Constancia M, Kelsey G, Reik W. Resourceful imprinting. Nature 2004; 432(7013):53-57.
23. Wilkins JF, Haig D. What good is genomic imprinting: The function of parent-specific gene expression. Nat Rev Genet 2003; 4(5):359-368.
24. Haig D. Genomic Imprinting and Kinship. New Brunswick, New Jersey: Rutgers University Press 2002.
25. Hamilton WD. The genetical evolution of social behaviour. J Theor Biol 1964; 7:1-16.
26. Haig D. Parental antagonism, relatedness asymmetries and genomic imprinting. Proc R Soc Lond B Biol Sci 1997; 264(1388):1657-1662.
27. Haig D. Placental hormones, genomic imprinting and maternal-fetal communication. J Evol Biol 1996; 9(3):357-380.
28. Ubeda F, Haig D. Dividing the child. Genetica 2003; 117(1):103-110.
29. Wilkins JF, Haig D. Genomic imprinting of two antagonistic loci. Proc R Soc Lond B Biol Sci 2001; 268(1479):1861-1867.
30. Tycko B, Morison IM. Physiological functions of imprinted genes. J Cell Physiol 2002; 192(3):245-258.
31. Killian JK, Nolan CM, Wylie AA et al. Divergent evolution in M6P/IGF2R imprinting from the Jurassic to the Quaternary. Hum Mol Genet 2001; 10(17):1721-1728.
32. Killian JK, Nolan CM, Stewart N et al. Monotreme IGF2 expression and ancestral origin of genomic imprinting. J Exp Zool 2001; 291(2):205-212.
33. Scott RJ, Spielman M. Genomic imprinting in plants and mammals: how life history constrains convergence. Cytogenet Genome Res 2006; 113(1-4):53-67.

34. Skuse DH, Purcell S, Daly MJ et al. What can studies on Turner syndrome tell us about the role of X-linked genes in social cognition? Am J Med Genet B Neuropsychiatr Genet 2004; 130B(1):8-9.
35. Plagge A, Isles AR, Gordon E et al. Imprinted Nesp55 influences behavioral reactivity to novel environments. Moll Cell Biol 2005; 25(8):3019-3026.
36. Plagge A, Gordon E, Dean W et al. The imprinted signaling protein XL alpha s is required for postnatal adaptation to feeding. Nat Genet 2004; 36(8):818-826.
37. Li LL, Keverne EB, Aparicio SA et al. Regulation of maternal care and offspring growth by paternally expressed Peg3. Science 1999; 284:330-333.
38. Lefebvre L, Viville S, Barton SC et al. Abnormal maternal behavior and growth retardation associated with loss of the imprinted gene Mest. Nat Genet 1998; 20:163-169.
39. Lawton BR, Sevigny L, Obergfell C et al. Allelic expression of IGF2 in live-bearing, matrotrophic fishes. Dev Genes Evol 2005; 215(4):207-212.
40. Crespi B, Semeniuk C. Parent-offspring conflict in the evolution of vertebrate reproductive mode. Am Nat 2004; 163(5):635-653.
41. Goodwin NB, Dulvy NK, Reynolds JD. Life-history correlates of the evolution of live bearing in fishes. Philos Trans R Soc Lond B Biol Sci 2002; 357(1419):259-267.
42. Reynolds JD, Goodwin NB, Freckleton RP. Evolutionary transitions in parental care and live bearing in vertebrates. Philos Trans R Soc Lond B Biol Sci 2002; 357(1419):269-281.
43. Haig D. Genetic Conflicts in Human Pregnancy. Quarterly Review of Biology 1993; 68(4):495-532.
44. Trivers R. Parent-offspring conflict. Am Zool 1974; 14:249-264.
45. Arngrimsson R. Epigenetics of hypertension in pregnancy. Nat Genet 2005; 37(5):460-461.
46. van Dijk M, Mulders J, Poutsma A et al. Maternal segregation of the Dutch preeclampsia locus at 10q22 with a new member of the winged helix gene family. Nat Genet 2005; 37(5):514-519.
47. Oudejans CBM, Mulders J, Lachmeijer AMA et al. The parent-of-origin effect of 10q22 in pre-eclamptic females coincides with two regions clustered for genes with down-regulated expression in androgenetic placentas. Mol Hum Reprod 2004; 10(8):589-598.
48. Haig D, Wharton R. Prader-Willi syndrome and the evolution of human childhood. Am J Hum Biol 2003; 15(3):320-329.
49. Wilkins JF, Haig D. Inbreeding, maternal care and genomic imprinting. J Theor Biol 2003; 221(4):559-564.
50. Brenton JD, Viville S, Surani MA. Genomic imprinting and cancer. Cancer Surv 1995; 25:161-171.
51. Jirtle RL. Genomic imprinting and cancer. Exp Cell Res 1999; 248(1):18-24.
52. Reik W, Constancia M, Fowden A et al. Regulation of supply and demand for maternal nutrients in mammals by imprinted genes. J Physiol 2003; 547(1):35-44.
53. Varmuza S, Mann M. Genomic Imprinting-Defusing the Ovarian Time Bomb. Trends Genet 1994; 10(4):118-123.
54. Wilkins JF. Genomic imprinting and methylation: epigenetic canalization and conflict. Trends Genet 2005; 21(6):356-365.
55. Gorelick R. Neo-Lamarckian medicine. Med Hypotheses 2004; 62(2):299-303.
56. Hanson MA, Gluckman PD. Developmental processes and the induction of cardiovascular function: conceptual aspects. J Physiol 2005; 565(1):27-34.
57. Waterland RA, Jirtle RL. Early nutrition, epigenetic changes at transposons and imprinted genes and enhanced susceptibility to adult chronic diseases. Nutrition 2004; 20(1):63-68.
58. Jablonka E, Lamb MJ. Epigenetic inheritance in evolution. J Evol Biol 1998; 11(2):159-183.
59. Gallou-Kabani C, Junien C. Nutritional epigenomics of metabolic syndrome-New perspective against the epidemic. Diabetes 2005; 54(7):1899-1906.
60. Hales NC, P. BDJ. The thrifty phenotype hypothesis. Br Med Bull 2001; 60:5-20.
61. Barker DJP. Fetal programming of coronary heart disease. Trends Endocrinol Metab 2002; 13(9):364-368.
62. Preece MA, Moore GE. Genomic imprinting, uniparental disomy and foetal growth. Trends Endocrinol Metab 2000; 11(7):270-275.
63. Haig D, Trivers R. The evolution of parental imprinting: a review of hypotheses. In: Ohlsson R, Hall K, Ritzen M, eds. Genomic Imprinting: Causes and Consequences. Cambridge, UK: Cambridge University Press, 1995:17-28.
64. Hurst LD, McVean GT. Growth effects of uniparental disomies and the conflict theory of genomic imprinting. Trends Genet 1997; 13(11):436-443.
65. Iwasa Y, Mochizuki A, Takeda Y. The evolution of genomic imprinting: Abortion and overshoot explain aberrations. Evol Ecol Res 1999; 1(2):129-150.
66. Burt A, Trivers RL. Genes in Conflict: The Biology of Selfish Genetic Elements. Cambridge: The Belknap Press, 2006.

CHAPTER 9

Evolutionary Theories of Imprinting— Enough Already!

Tom Moore* and Walter Mills

> *Again, for the naive falsificationist a theory is falsified by a "(fortified) observational" statement which conflicts with it (or rather, which he decides to interpret as conflicting with it). The sophisticated falsificationist regards a scientific theory T as falsified if and only if another theory T' has been proposed with the following characteristics: (1) T' has excess empirical content over T: that is, it predicts novel facts, that is, facts improbable in the light of, or even forbidden, by T, (2) T' explains the previous success of T, that is, all the unrefuted content of T is contained (within the limits of observational error) in the content of T'; and (3) some of the excess content of T' is corroborated.*
>
> —Imre Lakatos, *Falsification and the Methodology of Scientific Research Programmes* in *Criticism and the Growth of Knowledge.* Cambridge University Press, 1970:91-195.

Abstract

In our view, the conflict theory of imprinting explains the evolution of parental allele-specific gene expression patterns in the somatic tissues of mammals and angiosperms. Not surprisingly, given its importance in mammalian development and pathology, the evolution of imprinting continues to attract considerable interest from theoretical and experimental biologists. However, we contend that much of the ensuing debate is of poor quality. We discuss several problems with the manner in which workers in the field engage in this debate and we argue for a more formal approach to the discussion of theories of the evolution of imprinting.

Introduction

Much of what appears in the scientific literature is descriptive rather than analytical. This bias may be particularly prevalent in the biological sciences because of the inherent complexity and apparent arbitrariness of biological systems, presumably arising from the occurrence of random mutation and complex modes of sexual, kin and natural selection. The complexity of genomes, gene expression patterns, morphogenesis, ecology and animal behaviour means that Popper's scheme of hypothesis formulation and falsification, leading to theories with increasing explanatory and predictive power is rarely pursued explicitly. Rather, it seems to us that most experimental (and, in particular, developmental) biologists work to provide increasingly detailed descriptions of biological processes, which occasionally become generalized as laws.

A biological 'law' may be defined as a generalization from numerous observations. Exceptions to such laws are not fatal and may not be particularly uncommon. Biological laws are therefore "contingently true."[1] For example, there are many instances of breaches of Mendel's laws of inheritance of phenotypic traits among them, notably in the current context, genomic imprinting. However, it would be counterproductive to suggest that Mendel's laws are therefore rendered worthless for utilitarian purposes, such as genetic counselling or animal breeding.

*Corresponding Author: Tom Moore—Department of Biochemistry, Biosciences Institute, University College Cork, College Road, Cork, Ireland. Email: t.moore@ucc.ie, w_e_mills@yahoo.co.uk

Genomic Imprinting, edited by Jon F. Wilkins. ©2008 Landes Bioscience and Springer Science+Business Media.

In current scientific parlance a 'theory' may be characterised as an explanation for a particular problem or set of observations. The best current theory is the one that most parsimoniously accounts for the available data and there is an underlying assumption that all such theories are provisional. However, the way in which a new theory replaces the current one is open to considerable debate. Popper's *falsificationist* methodology,[2] which is well known in the scientific community, has been criticised as being too "logically neat"[3] and alternative scenarios have been proposed that perhaps better reflect the way scientists and research programs actually work.[4,5] However, even Popper accepted that well-founded theories are rarely rejected on the basis of a single or an isolated set of conflicting observations. Rather, different theories compete with one another to provide an explanation for the available data and the most parsimonious theory is adopted. However, observing that there are exceptions or incompatibilities between the current theory and the available data is of little value if it does not motivate refinement of the theory or its replacement with a better one.

In our view a law, unlike a theory, does not necessarily rest on knowledge or understanding of the underlying mechanisms. For example, Mendel knew nothing of the chemical nature of the gene or the cellular basis of meiosis when formulating his laws: he constructed them purely on the basis of empirical observations. Only much later was it understood that the mechanistic basis of his laws lay in the random segregation and recombination of chromosomes. When we ask under what selection pressures sex, recombination and meiosis might have evolved we find that there are many well-founded theories, none of which, however, has gained pre-eminence because none of them appears to explain all of the empirical data (e.g., refs. 6-9).

Compared to the long history of studies of genetic recombination and meiosis, the history of mammalian imprinting is relatively compressed. There has been dramatic progress in identifying imprinted genes and in elucidating imprinting mechanisms over the last approximately twenty years since the first major descriptions of the process.[10-12] As with studies of meiosis, there is now general agreement regarding mechanisms. For example, the involvement of DNA methylation is well established.[13,14] Even so, much of the detail remains to be elucidated. Although no 'laws' of mammalian imprinting have been explicitly defined, there are nevertheless features that could be so classified; for example, paternally expressed imprinted genes promote fetal and placental growth, whereas maternally expressed genes do the opposite.[12] As in the case of Mendel's laws, this statement can be made on the basis of empirical observation alone and requires no knowledge of the underlying mechanisms or selective forces. It can, however, guide us towards a viable adaptive theory.

In contrast to our relatively incomplete understanding of the evolution of recombination and meiosis, we argue that there is a single pre-eminent theory that identifies the selective forces leading to the evolution of parental imprinting, namely, the conflict theory.[15,16] We contend that the available evidence strongly corroborates this theory and, moreover, that it is 'risky' in the Popperian sense: it makes specific predictions that are open to refutation. For example, the discovery of parental imprinting of genes expressed in the somatic tissues of a species that does not engage in parental investment would seriously undermine the theory. However, perusal of the literature on imprinting illustrates that a significant number of workers in the field either do not accept the pre-eminence of the conflict theory, or appear to be ignorant of the formal methods of introducing new theories. In particular, we contend that a significant proportion of the models of imprinting produced by mathematical biologists are of limited value and merely confuse the debate. In this article we highlight some of these problems and attempt to show why the conflict theory should be regarded as the best currently available explanation of the evolution of imprinting.

What Needs to be Explained?

Does Imprinting Require an Adaptive Explanation?

Evolutionary theories are primarily concerned with elucidating the identity and dynamics of the selective forces that result in the evolution of adaptive complexity. However, before discussing the various theories of evolution of imprinting, we must first ask whether imprinting actually demands an adaptive explanation i.e., one that assumes it to be selectively advantageous, or whether it is merely an epiphenomenon? We note that the 'null hypothesis' against which all adaptive theories must be tested is the possibility that the spread of a new allele and its associated phenotype may have occurred by genetic drift or by hitchhiking on a genetically linked advantageous variant. Of course, particularly at a gross level, the adaptive value of a phenotype may be self-evident: legs are for locomotion etc. However, at finer levels of organismal structure or behavior, the relevant selective forces may be more difficult to discern. One general refutation of drift and hitchhiking as explanations would be to show that there are significant costs associated with the relevant phenomenon. In the case of imprinting it has been proposed that there may be a cost associated with monoallelic expression due to the exposure of deleterious recessives.[17] We could then argue that imprinting must indeed be advantageous in order to counteract such costs. Similarly, the distribution or conservation of a phenomenon in phylogeny or ontogeny might argue in favour of an adaptive function. Finally, the phenomenon may be predicted by theory. We discuss these points below.

Does Imprinting Evolve in Opposition to the Evolutionary Costs of Monoallelic Expression?

Among complex multicellular organisms the haploid condition is generally confined to cell lineages that produce the germ cells. Therefore, diploidy (or higher polyploid states) is generally the norm in the somatic tissues of vertebrates and plants. The advantages conferred by diploidy are uncertain; however, its widespread occurrence in phylogeny strongly suggests that it is advantageous and credible explanations have been proposed: for example, diploidy may protect against either inherited or somatic mutations (see ref. 18 for a discussion). Instances of monoallelic expression of individual diploid genetic loci are therefore exceptional and may require an explanation. However, there are two reasons why monoallelic expression at a small number of genetic loci may not be particularly costly. First, as noted by Spencer,[19] at equilibrium there is one selective death for each inherited deleterious mutation, therefore selection against monoallelic expression would be of the same order of magnitude as the mutation rate. Given the small number of imprinted genes, this is unlikely to be a significant cost. Second, the stoichiometry of gene products can, in some cases, be a critical determinant of the phenotype at dosage sensitive genetic loci.[20,21] At biallelically expressed loci of this type, mutation of one copy of the gene may be deleterious and the advantage of diploidy may be reduced or non-existent. If such a locus evolves monoallelic expression due to imprinting, there may be no significant additional cost associated with the imprinted state relative to the biallelic state. Indeed, the conflict theory predicts that it is precisely at such dosage sensitive loci that imprinting evolves.[22] Therefore the significance of the putative costs due to monoallelic expression at imprinted loci may have been overstated. For a further discussion of the selective consequences of functional haploidy at imprinted loci, see the chapter by Úbeda and Wilkins.

The Restricted Phylogeny of Imprinting

Differential expression of the parental alleles at a small number of loci has been described in the somatic tissues of mammals and in the endosperm of angiosperm plants.[23,24] However, in spite of extensive genetic analyses, no evidence of imprinting of endogenous genes has been discovered in the somatic tissues of other phylogenetic groups such as birds,[25,26] fish,[27] flies,[28] or worms.[29] The conflict theory provides a neat explanation for this distribution because mammals and endosperm plants exhibit the two key features it holds to be required for imprinting to evolve: first, alleles in progeny are able to manipulate the level of maternal investment and, second, polyandry reduces the genetic relatedness of paternal alleles relative to maternal alleles in progeny.[16] We note that mathematical models suggest that even fractional degrees of polyandry (e.g., due to death and replacement of mates) may be sufficient for imprinting to evolve.[22] We note also that, under conflict, imprinting could evolve in oviparous species such as birds; for example, through manipulation of begging behaviour of nestlings in altricial species. However, no evidence of this has been uncovered. The conflict theory does not claim to explain the imprinting and inactivation of whole chromosomes or sets of chromosomes in insects such as *Sciaridae*, *Scolytidae* and *Pseudococcus* species. However, a broader version of kinship theory may well do so.[31,40]

Key Features of Imprinted Genes

Current estimates suggest that less than one percent of genes are imprinted in mammals.[24] This can be explained under the conflict theory because only genes that are both dosage sensitive and expressed in tissues that influence maternal investment will be selected.[33] We know from classical Mendelian genetics and engineered mouse null mutants that only a small proportion of genes exhibit detectable dosage sensitivity. This greatly reduces the pool of genes that are candidates for imprinted expression.

The approximately eighty imprinted genes identified in the mouse and human are heterogeneous with respect to their biochemical and cellular functions.[24] Nevertheless, the expression patterns of imprinted genes and the phenotypes of mice and humans resulting from their deregulation offer important clues to their developmental functions. First, they are predominantly expressed at embryonic and early postnatal stages of development.[12] Second, there is a preponderance of abnormal growth and behavioral phenotypes in mutants.[12] Third, a large proportion of imprinted genes are expressed in the placenta and influence its growth, development and function.[34] Fourth (and critically for the conflict theory), maternally and paternally inherited imprinted alleles tend to exhibit opposite effects on growth and behavioural parameters.[12] Analogous observations have been made with respect to endosperm growth in plants.[15,35] As has been argued previously, these empirical data strongly corroborate the conflict theory because the major routes by which offspring solicit maternal investment are the placenta and suckling in mammals and the endosperm in plants.[16] Finally, we note that the process of paternal X chromosome imprinting and inactivation in mammals can also be accommodated within the conflict theory.[36,37]

Can a Theory Supersede a Fact?

Huxley's aphorism—"A beautiful theory slain by an ugly fact"—exquisitely encapsulates the view that the fact should reign supreme in science. We assume that by 'fact', Huxley meant primary data or observations. However, 'facts' are themselves predicated on underlying theories that are open to refutation.[2] Particularly in the complex and technically challenging areas of molecular genetics and developmental biology, facts may indeed be provisional; for example, there are numerous instances where the observed function of a gene depends on the genetic background or the physiological context in which its effects are measured. As we noted earlier, isolated examples of imprinted genes, such as *Mash2*, that appear to contradict the conflict theory should be viewed in this light until auxiliary hypotheses (see below) are developed that explain the anomaly, or until there is an accumulation of examples of such genes so as to seriously question the validity of the theory. We further suggest that, given the provisional nature of many observations in molecular and developmental genetics, the explanatory and predictive power of a mathematically grounded theory such as kin selection may be such that it supersedes at least some of the facts. Or, as Einstein allegedly had it: "If the facts don't fit the theory, change the facts." In other words, incompatible data should prompt a re-examination of the observation rather than the immediate abandonment of the theory. We note that this prescription leads to a recursive process of data selection and theory reinforcement. This process will naturally delimit the explanatory boundary of a theory, so that if it is applied too widely, falsifying observations begin to accumulate. The danger is that it may lead to a proliferation of proprietary mini-theories whereby each worker develops a theory founded exclusively on the features of their preferred imprinted gene (e.g., ref. 38).

The Etiquette of Proposing a New Theory of Imprinting

The principle of parsimony requires that the simplest theory (with the fewest assumptions or logical steps) should be adopted in preference to more complex explanations of the same problem. However, before adjudicating between competing theories, we must decide whether the theories actually claim to explain the same data. As noted above, there may be disagreement over precisely which data are to be considered and this may impact on which theory is the most parsimonious. For example, many molecular and developmental biologists who comment on the evolution of imprinting habitually ignore imprinting in the plant endosperm (e.g., ref. 39). Others consider autosomal imprinting in mammals and plants and chromosomal imprinting in insects (e.g., ref. 40) or consider mammalian X chromosome imprinting exclusively (e.g., refs. 41, 42). There is no general solution to this problem and workers in the field must either arrive at a consensus, or agree to disagree, or simply disagree. The conflict theory claims to explain both autosomal and X chromosomal imprinting in the somatic tissues of mammals and plants.[15,16] Compared to other attempts to explain the evolution of imprinting, it is therefore highly inclusive with respect to the data considered.

Once a coherent problem or set of observations is agreed upon, a hypothesis or theory may be tested against the data to determine whether it is corroborated or falsified. A substantial, well-founded theory should receive significant corroboration from the data and its main features should resist falsification. However, as noted above, a limited degree of conflict with the data is tolerable because such exceptions may be rationalized using auxiliary hypotheses.[5] A good example with respect to the conflict theory is the observation that a null mutant of the maternally expressed mouse *Mash2* gene results in failure of placental development,[43] contrary to the expectation that ablation of maternally expressed imprinted genes should cause placental overgrowth or other abnormalities consistent with pathologically increased maternal investment in the embryo. Initially, this observation led some commentators to claim that the conflict theory had been falsified (see ref. 44 for discussion). However, there are several hypotheses that are consistent with the conflict theory that could account for this observation. For example, recent evidence indicates that *Mash2* promotes survival of trophoblast progenitor cells, which is consistent with the predictions of the conflict theory.[45] The most likely explanation, in our view, is that the *Mash2* null mouse phenotype, by virtue of its severity, may not be particularly informative about *Mash2* function in normal pregnancy. Subtle mutations, in which *Mash2* gene dosage is incrementally increased or decreased, may be the most suitable experimental system for testing the conflict theory, albeit prohibitively expensive.

Since the publication of the conflict theory,[15,16] there have been numerous alternative proposals to explain imprinting. The merits (or lack thereof) of the conflict theory and many of the alternative proposals have been discussed extensively elsewhere and will not be repeated here.[46,47] We note, however, that the comprehensive and repeated rebuttals notwithstanding, some alternative proposals are still alluded to as if they were viable alternatives to the conflict theory (e.g., refs. 10 and 24). The confusion produced by the proliferation of poorly formulated theories of imprinting, which are either uncorroborated or poorly corroborated by the empirical data, is further increased when mathematical models endow them with a false sheen of rigour. For example, the 'ovarian time bomb' (OTB) hypothesis has been modelled mathematically by Weisstein et al,[48] who show that, under the assumptions of their model, it can predict the directionality of phenotypes associated with maternally and paternally silenced imprinted genes. However, it was already clear that there were numerous aspects of the data that the OTB hypothesis could not explain, but which are explicable under the conflict theory.[47] Parsimony should therefore have dictated that the OTB be rejected in favour of the conflict theory, without troubling the mathematicians. Similarly, Weisstein and Spencer[49] modelled the rather obscure hypothesis of 'variance minimization' even though it has virtually no explanatory power with respect to the empirical data. Day and Bonduriansky[50] produced a mathematical model of their hypothesis that imprinting evolved through intralocus sexual conflict. The authors identify a selective force with "very broad applicability" in phylogeny. However, as the empirical data clearly indicate that imprinting is phylogenetically highly restricted, this can hardly count as support for their hypothesis. Moreover, in their discussion they suggest that their hypothesis and the conflict theory may not be mutually exclusive. However, as both proposals lay claim to some of the same data (e.g., imprinting of *Igf2*), this is not a parsimonious solution. The parsimony principle must also be focussed on the proposal that imprinting of X-linked genes may have been selected to promote sexual dimorphism in mammals.[41,42] Again, the fact that the more inclusive conflict theory can explain the same data, without appealing to an additional selection pressure, should lead to rejection of the sexual dimorphism hypothesis.[37]

Conclusion

Since the first comprehensive descriptions of mammalian imprinting in the mid-1980's the topic has continued to fascinate geneticists, developmental biologists, molecular pathologists and evolutionary biologists. Indeed, the field of imprinting has been at the vanguard of important areas of molecular genetics such as chromatin regulation, noncoding RNA and cancer epigenetics. It is entirely natural that researchers should feel a degree of ownership of a phenomenon to which they have devoted considerable effort. Moreover, it behoves theoreticians to properly master the empirical facts before speculating on the evolution of imprinting. Conversely, researchers should be prepared to accept that, apart from delineating the empirical facts, they may not be best placed to solve the problem of the evolution of imprinting.

We believe that the conflict theory is the pre-eminent explanation of the evolution of imprinting. In the early days, the theory received a sympathetic hearing from researchers. However, we suggest that there has been a considerable amount of unproductive commentary from (predominantly) mathematical biologists who have not taken sufficient trouble to master the empirical data (or, indeed, the scientific method). In extreme cases, the primary aim appears to be the identification of data (any data!) that provide an opportunity to flaunt one's mathematical skills, rather than to solve a biological problem per se. The triumphant mathematician can then, like a caricature of Oliver Goldsmith's village school master, accept due admiration from the pedestrian researchers ("While words of learned length and thundering sound/Amazed the gazing rustics ranged around"). Enough already!

References

1. Beatty J. The evolutionary contingency thesis. In: Wolters G, Lennox JG, McLaughlin P, eds. Concepts, Theories and Rationality in the Biological Sciences: the Second Pittsburgh-Konstanz Colloquium in the Philosophy of Science. Pittsburgh: University of Pittsburgh Press, 45-81.
2. Popper KR. The Logic of Scientific Discovery. New York: Harper. 1959.
3. Thornton S. "Karl Popper", The Stanford Encyclopedia of Philosophy (Summer 2005 Edition), Zalta EN(ed.), URL = <http://plato.stanford.edu/archives/sum2005/entries/popper/>.
4. Kuhn TS. The Structure of Scientific Revolutions Chicago:University of Chicago Press, 1962.
5. Lakatos I. Falsification and the Methodology of Scientific Research Programmes. In: Lakatos I, Musgrave A, eds. Criticism and the Growth of Knowledge. Cambridge: Cambridge University Press. 1970:91-195.
6. Maynard Smith J. The Evolution of Sex. Cambridge: Cambridge University Press. 1978.
7. Hamilton WD. Sex versus nonsex versus parasite. Oikos 1980; 35:282-290.
8. Kondrashov AS. Deleterious mutations and the evolution of sexual reproduction. Nature 1988; 336:435-440.
9. Peters AD, Otto SP. Liberating genetic variance through sex. Bioessays 2003; 25(6):533-7.
10. Miyoshi N, Barton SC, Kaneda M et al. The continued quest to comprehend genomic imprinting. Cytogenet Genome Res 2006; 113(1-4):6-11.
11. Solter D.Imprinting today: end of the beginning or beginning of the end? Cytogenet Genome Res 2006; 113(1-4):12-6.
12. Cattanach BM, Beechey CV, Peters J. Interactions between imprinting effects: summary and review. Cytogenet Genome Res 2006; 113(1-4):17-23.
13. Sasaki H, Ishihara K, Kato R. Mechanisms of Igf2/H19 imprinting: DNA methylation, chromatin and long-distance gene regulation. J Biochem (Tokyo) 2000; 127(5):711-5.
14. Holmes R, Soloway PD. Regulation of imprinted DNA methylation. Cytogenet Genome Res 2006; 113(1-4):122-9.
15. Haig D, Westoby M. Parent-specific gene expression and the triploid endosperm. American Naturalist 1989; 134:147-155.
16. Moore T, Haig D. Genomic imprinting in mammalian development: a parental tug-of-war. Trends in Genetics 1991; 7:45-49.
17. Mochizuki A, Takeda Y, Iwasa Y. The evolution of genomic imprinting. Genetics 1996; 144(3):1283-95.
18. Orr HA. Somatic mutation favors the evolution of diploidy. Genetics 1995; 139(3):1441-7.
19. Spencer HG. Mutation-selection balance under genomic imprinting at an autosomal locus. Genetics 1997; 147(1):281-7.

20. Wray GA. Transcriptional regulation and the evolution of development. Int J Dev Biol 2003; 47(7-8):675-84.
21. Pritchard C, Coil D, Hawley S et al. The contributions of normal variation and genetic background to mammalian gene expression. Genome Biol 2006; 7(3):R26.
22. Mills W, Moore T. Polyandry, life-history trade-offs and the evolution of imprinting at Mendelian loci. Genetics. 2004 168(4):2317-27. Erratum in: Genetics 2005; 171(3):1443.
23. Vinkenoog R, Bushell C, Spielman M et al. Genomic imprinting and endosperm development in flowering plants. Mol Biotechnol 2003; 25(2):149-84.
24. Morison IM, Ramsay JP, Spencer HG. Trends Genet 2005; 21(8):457-65.
25. O'Neill MJ, Ingram RS, Vrana PB et al. Allelic expression of IGF2 in marsupials and birds. Dev Genes Evol 2000; 210(1):18-20.
26. Nolan CM, Killian JK, Petitte JN et al. Imprint status of M6P/IGF2R and IGF2 in chickens. Dev Genes Evol 2001; 211(4):179-83.
27. Lawton BR, Sevigny L, Obergfell C et al. Allelic expression of IGF2 in live-bearing, matrotrophic fishes. Dev Genes Evol 2005; 215(4):207-12.
28. Wittkopp PJ, Haerum BK, Clark AG. Parent-of-Origin Effects on mRNA Expression in Drosophila melanogaster Not Caused by Genomic Imprinting. Genetics 2006; 173(3):1817-21.
29. Haack H, Hodgkin J. Tests for parental imprinting in the nematode Caenorhabditis elegans. Mol Gen Genet 1991; 228(3):482-5.
30. Goday C, Esteban MR. Chromosome elimination in sciarid flies. Bioessays 2001; 23(3):242-50.
31. Queller DC. Theory of genomic imprinting conflict in social insects. BMC Evol Biol 2003; 3:15.
32. Khosla S, Mendiratta G, Brahmachari V. Genomic imprinting in the mealybugs. Cytogenet Genome Res 2006; 113(1-4):41-52.
33. Moore T. Genetic conflict, genomic imprinting and establishment of the epigenotype in relation to growth. Reproduction 2001; 122(2):185-93.
34. Fowden AL, Sibley C, Reik W et al. Imprinted genes, placental development and fetal growth. Horm Res 2006; 65 Suppl 3:50-8.
35. Lin B-Y. Association of endosperm reduction with parental imprinting in maize. Genetics 1982; 100:475-486.
36. Moore T, Hurst LD, Reik W. Genetic conflict and evolution of mammalian X chromosome inactivation. Dev Genet 1995; 17(3):206-11.
37. Mills, W and Moore, T. Evolution of mammalian X chromosome-linked imprinting. Cytogenet Genome Res 2006; 113(1-4):336-44.
38. Okamura K, Ito T. Lessons from comparative analysis of species-specific imprinted genes. Cytogenet Genome Res 2006; 113(1-4):159-64.
39. Kaneko-Ishino T, Kohda T, Ono R et al. Complementation hypothesis: the necessity of a monoallelic gene expression mechanism in mammalian development. Cytogenet Genome Res 2006; 113(1-4):24-30.
40. Normark BB. Perspective: maternal kin groups and the origins of asymmetric genetic systems-genomic imprinting, haplodiploidy and parthenogenesis. Evolution Int J Org Evolution 2006; 60(4):631-42.
41. Skuse DH. Genomic imprinting of the X chromosome: a novel mechanism for the evolution of sexual dimorphism. J Lab Clin Med 1999; 133(1):23-32.
42. Iwasa Y, Pomiankowski A. The evolution of X-linked genomic imprinting. Genetics 2001; 158(4):1801-9.
43. Guillemot F, Caspary T, Tilghman SM et al. Genomic imprinting of Mash2, a mouse gene required for trophoblast development. Nat Genet 1995; 9(3):235-42.
44. Iwasa Y. The conflict theory of genomic imprinting: how much can be explained? Curr Top Dev Biol 1998; 40:255-93.
45. Ferguson-Smith AC, Moore T, Detmar J et al. Epigenetics and imprinting of the trophoblast—a workshop report. Placenta 2006; 27 Suppl A: S122-6.
46. Hurst, LD. Evolutionary theories of genomic imprinting. In: Reik W, Surani A, eds. Genomic Imprinting. Oxford: IRL Press, 1997; 211-37.
47. Wilkins JF, Haig D. What good is genomic imprinting: the function of parent-specific gene expression. Nature Reviews Genetics 2003; 4:359-368.
48. Weisstein A, Spencer HG. Evolutionary Genetic Models of the Ovarian Time Bomb Hypothesis of Genomic Imprinting. Genetics 2002; 162:425-439.
49. Weisstein A, Spencer HG. The Evolution of Genomic Imprinting via Variance Minimization: An Evolutionary Genetic Model. Genetics 2003; 165:205-222.
50. Day T, Bonduriansky R. Intralocus sexual conflict can drive the evolution of genomic imprinting. Genetics 2004; 167(4):1537-46.

INDEX

Symbols

1,25-dihydroxy vitamin D3 28

A

Adaptation 41, 53, 67
Adipocyte 44, 45, 48, 53, 54, 56
Adipocyte differentiation 45, 56
Adipose tissue 17, 18, 31, 32, 41, 43-46, 48, 50, 53-56
Albright's hereditary osteodystrophy (AHO) 27, 29-32, 41, 54
Alternative splicing 47
Androgenetic 43, 64, 72, 90
Angelman syndrome (AS) 41-43, 55, 63-65, 71, 76, 104, 107, 108, 113
Antisense transcript 16, 24, 35, 63
Arabidopsis 90-94, 96
Asperger's syndrome 74, 76
Atp10c 43, 46, 56
Attention Deficit/Hyperactivity disorder (ADHD) 71, 79, 80
Autism 63, 65, 68, 71, 74, 80, 106
Autistic Spectrum disorder (AD) 27, 32-35, 74-76
Autosomes 10

B

Behavior 43, 62, 65-68, 71-81, 103, 104, 106-108, 113, 116, 118
Brain 18, 42-45, 47, 53, 62-68, 71-75, 77-80, 107, 108
Brown adipose tissue 44
Brown fat 21, 22, 24

C

Cancer 1, 2, 78, 80, 105, 108, 113, 121
Cell-specific imprinting 63
Chimera 64, 72, 80
Chromatin 12, 19, 20, 22, 24, 63, 89, 94, 96, 101, 111, 121
cis-acting imprinting control element 17
Cognition 65, 71-74, 81

Complex *GNAS* locus 36
Conduct disorder (CD) 79, 80
Conflict hypothesis 55, 90
Conflict theory 56, 67, 89, 90, 116-121
CpG island 1, 3, 11, 12, 19, 20
Cyclic AMP (cAMP) 18, 19, 27-31, 46, 49, 53, 54

D

Demeter (DME) 5, 92, 93, 95, 96
Demethylation 3, 5-8, 10-12, 45
Developmental origins of adult disease 66
Differentially methylated domain (DMD) 93, 94
Differentially methylated region (DMRs) 3-5, 18, 20, 22, 23, 27, 32, 34, 36, 47, 78
DNA methylation 1, 3, 9-12, 20, 89, 92-94, 96, 111, 117
Domains rearranged methyltransferase (DRM) 94

E

Empathy 73
End-organ resistance 27, 36
Endosperm 89-96, 118, 119
Energy homeostasis 41, 44-46, 48, 49, 53, 54, 56
Epigenetics 5, 9, 11, 12, 16, 27, 29, 33-36, 56, 62, 66, 78, 80, 89, 92, 94-96, 101, 109, 111-113, 121
Epigenotype 101
Epimutation 94, 109, 111-113
Evolution 19, 47, 73-76, 89, 90, 96, 101-103, 108, 116, 117, 119-121
Evolution of genomic imprinting 89
Exon 1A 17-24, 36, 47
Exon A/B 20, 27, 32-36
Extra-large α-stimulatory G protein (XLαs) 18, 19, 23, 24, 46, 47, 49, 50, 53

F

Falsification 116, 120
Feeding 42, 43, 49, 53, 55, 65, 72, 76-78, 101, 104, 113
Fertilisation independent endosperm 1 (FIE1) 94
Fertilisation independent seed 2 (FIS2) 90, 92, 93, 95
Fetal programming 56, 78, 112
Functional haploidy 108, 109, 111, 118
FWA 92-96

G

Gamete 1, 8, 10, 11, 91, 94
Gene therapy 113
Genomic imprinting 16, 55, 56, 62, 71-73, 76, 78, 80, 89-91, 93-96, 101-103, 109, 112, 116
GNAS 16-24, 27-36, 41, 46-56, 67
GNAS antisense transcript 35
Gnasxl 18-23, 46-48, 53-56, 64, 65
G protein 27
G protein-coupled receptor 27
Gsα 18, 19, 21, 23, 24, 47-51, 53, 54
Gynogenetic 72, 90

H

Heritability 73
Hippocampus 43, 63, 66, 72
Human placental lactogen (hPL) 107
Hyperphosphatemia 31
Hypocalcemia 28, 31
Hypothalamus 42-44, 53, 64, 72, 75, 77, 78

I

IGF2 2, 4, 5, 12, 22, 45, 46, 63, 66, 77, 78, 90, 94, 120
Imprinting 1, 4, 10, 12, 16, 17, 19, 20, 22-24, 27, 28, 30, 32-36, 41-43, 45, 46, 48, 53-56, 62-64, 66-68, 71-80, 89-96, 101-103, 107-109, 111-113, 116-121
Imprinting center 19, 20
Imprinting control region (ICR) 1-3, 9, 11, 12, 19-23, 93, 94
Inbreeding 55, 108
Inclusive fitness 73, 102, 107, 109, 112

K

Kin selection 55, 119
Kinship theory 19, 55, 67, 74, 102, 107, 108, 118
Knockout mice 5, 31, 48, 56, 64-67

M

Maize 90, 91, 94
Maternal care 66, 103, 104, 107
Maternal duplication 63
Medea (MEA) 90-96
Mesoderm-specific transcript (*Mest*) 9, 43-46, 56, 65, 66, 108 *see also* Paternal expressed gene 1
Metabolic syndrome 111, 112
Metabolism 31, 41, 43-49, 53-56, 71, 72, 76-78, 80, 111
microRNA 18
Mood disorder 79

N

NDN 42, 45, 56, 64, 66, 105
Necdin 45, 64
Neocortex 64, 72, 73, 75
Nesp 18, 20-23, 47, 64, 66, 67
NESP55 (55 KD neuroendocrine secretory protein) 20, 27, 29, 35, 36
Nespas 17, 18, 20-24, 47
Neurodevelopment 64, 66, 67
Neuroendocrine 18, 43, 45, 55, 75
Neurological 62, 63, 79
Neuropsychiatric 62, 63, 71, 76
Noncoding RNA 20, 23, 101, 111, 121

O

Obesity 29, 31, 41-43, 45, 48, 54-56, 76-79, 105, 113
Obsessive-compulsive disorder (OCD) 78, 79
Oed-Sml 19
Ovarian time bomb (OTB) 120
Oxytocin 64, 68, 72, 79

P

Parathyroid hormone (PTH) 20, 21, 27-29, 31-36, 47, 53, 104
Parent-of-origin effect 11, 43, 63, 68, 72-74, 77, 80, 81, 90, 91
Parental conflict 75, 89, 90, 103
Parthenogenetic 9, 43, 64, 72, 90
Paternal expressed gene 1 (*Peg1*) 9, 43-46, 56, 65, 66, 108 *see also* Mesoderm-specific transcript
Paternal expressed gene 3 (*Peg3*) 4, 9, 43, 44, 46, 55, 56, 64-66, 68, 108
PHERES1 92, 93, 95, 96
Placenta 44, 66, 89, 94, 105, 107, 119
Polycomb 92, 95
Polycomb Repressive Complex 2 (PRC2) 92, 93, 95, 96
Postnatal metabolism 41, 55, 76
Prader-Willi syndrome (PWS) 21, 23, 41, 42, 45, 55, 56, 63-66, 71, 76-80, 105, 107, 108
Pre-eclampsia 104, 107
Pref1/Dlk1 (Preadipocyte factor 1/Delta, Drosophila, Homolog-like 1) 45, 46, 55, 56
Principle of parsimony 119
Progressive osseous heteroplasia (POH) 29, 30
Pseudohypoparathyroidism (PHP) 20, 27-36, 54, 104
Pseudopseudohypoparathyroidism (PPHP) 27, 29-32
Psychiatry 75
Psychopathology 72, 78
Psychosis 65

R

Repeat 1-5, 8-12, 33, 34, 45, 93, 94, 96
Reprogramming 1, 3, 5, 11, 111, 112

S

Satellite 2, 4, 9, 11
Scientific method 121
Septum 43
Serotonin 77, 79, 80
Sex-biased dispersal 67
Social behavior 72-75, 103
STOX1 104, 107
STX16 17, 20, 27, 29, 33-35
Suckling 19, 44, 49, 53, 65, 77, 107, 108, 119
Syntaxin-16 34

T

Theory 12, 19, 41, 55, 56, 62, 67, 73-75, 89, 90, 102, 107, 108, 116-121
Theory of mind 73-75
Tissue-specific silencing 35, 36
Transposon 3
Turner's syndrome (TS) 65, 74

U

Ube3a 23, 43, 63-65, 67
Uncoupling protein 1 (UCP1) 45, 48, 53

V

Viviparity 103

W

Waddington, C.H. 101

DATE DUE

DUE DATE SUBJECT TO CHANGE
IF A RECALL IS REQUESTED